Cavalry Tales

Cavalry Tales. Copyright © 2020 by Carroll Childers, MG (USA Retired) and C Troop et al. All rights reserved. This book or any portion thereof may not be reproduced or used in any manner whatsoever without the express written permission of the publisher except for the use of brief quotations or citations in scholarly works, review, etc.

Printed in the United States of America

First Printing, 2020

ISBN 978-1-7353222-0-9

Publisher
Carentan Media Group
2215 Plank Rd #165
Fredericksburg, VA. 22401

FOREWARD

It is not clear who may have first uttered some not so well thought out statement which has now become Cavalry Tales. It is clear that the product was the result of activity by many and these we credit here by name and contribution.

- CONTRIBUTING AUTHORS : Carroll Childers, L.P. Hening, Tony Burke, Mike Thomas, Bryce Bugg, Alan Huhes, Neil Hening, Bob Bruce, Ed Heavener, Bill Talbott, John Terry, Roddy Davoud, Bob Anderson, Rick Meador, Tom Redfern,
- Donors of photographs Bill Talbott, Bill Atkeison, L.P. Hening, Bob Bruce, and Amir Pishdad
- Coordination of files --Mike Thomas
- Those with stories they can't author it so a Transcriber was needed – Mike Thomas
- Squeezing heads for stories – Mike Thomas
- Sanitization of text -- Mike Thomas and Carroll Childers (the delete supply is now low)
- Final Edit and Layout by - Jim Bartlett

Troop C is thrilled to offer their deepest expression of appreciation to Trooper Roddy Davoud who graciously and unconditionally provided the resources to:

- Generate and produce this historical artifact.
- Establish a critical tool (Professional website) to maintain comms among the Troopers.
- And especially for a grand and momentous annual C-Troop reunion hosted at Farview by Roddy and Cindi Davoud where Troopers and their significant others, and special invited guests are welcomed, entertained lavishly, and long term friendships are renewed.

For more info on C Troop, 223rd Cavalry visit:
www.ctroop223.org

Table of Contents

Introduction
C Troop Honor Roll

Basic History of the VA Army National Guard and Armored Cavalry	1
Alfred "Bull" Brammer	5
Always Eat Last	6
And Then There Was The Time.....	8
AT-73 Camp Pickett / AWOL Trooper	12
Baileyisums	17
Blow the Field Latrine	18
Camaraderie	22
C Troop Call Ups	24
C Troop Memories	30
C Troop movie Time	33
Cornfield Detour	36
VAARNG OCS	40
Tents Galore - Ft Drum	45
Reminisce – COL Ed Heavener	46
The Human Flame Thrower Affair	49
Getting Fired By the Best	51
Grounding the Aviation Fleet	57
Guarding the Mess Hall	74
IG Inspection	76
A Knife at a Poker Game	78
Whoops!	79
Incoming! Outside the Lines at Ft. Pickett	81

Integrity	85
Joining the Guard in AR	87
Whistlers	92
Legends, Rumors, and Myths	93
The M-60 MG Bandit	100
Meeting SFC Macon Morris	101
My Interactions with various TAGs	104
Never Forget Your Friends	116
Officer Candidate School (OCS)	118
Platoon Level Crew Drill	123
The Porcupine Incursion	128
Possum Is Not Meat at All	129
Push Up Challenge	130
Riding the Huey Skid	144
Supply Officer Turn In	148
Tank Incident Nottoway R. Bridge	149
Tank Rescue	154
Tank Section Training Accident	160
The Drive Home	165
The Pleasure of Armored Cavalry	167
Top of the Heap	171
Thundering Herd	182
Wash Point Woodsey	183
The Only Command out of reach	184
Life in the Cavalry	189
Flood Duty	195
Tank Commander	197

Poker Anyone	198
Soldier-You Owe the USA $87K	199
Another Invoice for Talbot	201
Golf Anyone	202
Lead, Follow, or Get Out Of The Way	204
NRDUV Letter of Commendation	208

Appendices: C Troop Commander History

Introduction

Only the soldier who has been a member of an Armored Cavalry Troop (ACT) can fully appreciate the great personal reward that one derives from accomplishing the mission assigned. Being a Trooper is difficult and dangerous, even in a training environment. It is characterized by long training days calling for considerable strength and conditioning in the performance of the individual and team tasks. Operating 100 percent of the time in the environment that God provides, exposed to the elements, it requires being mentally alert in an environment which can result in serious injury and being reflexive to these mechanisms of injury. Every move that a soldier makes creates a chain of events which may impact a fellow soldier.

Take the standard task of driving a tank, wherein the skill and caution taken by the driver can impact any or all of the other three crewmen as they are pitched and thrown about at their crew stations in response to how the tank is maneuvered. Tanks may actually find themselves out of contact with the earth and be temporarily airborne before returning to earth with a bone jarring transfer of forcc to the crew. The bulk of the ACT rides on Full Tracked Armored Vehicles where speed is life. The remainder of the ACT rides on wheeled vehicles but again, speed is life on the battlefield and one must train as they expect to fight.

Keep in mind that the subject of this book is the ACT Organization and Equipment of the era 1964-1985, in the Army National Guard, and not to be confused with the 21st Century configuration of similar units.

Another characteristic of the ACT is that because they train hard and realistically in a tough and challenging environment, a strong bond of Camaraderie is formed which, at least for C Troop 223rd Armored Cavalry Squadron, has survived for decades. As C Troop began to revive old ties, some three decades after the unit was officially disbanded, having hosted 6 C Troop commanders and as many First Sergeants and an uncounted number of Platoon leaders and Sergeants, a common desire began to be discussed to resurrect and record the experiences of the ACT for posterity and for the benefit of families who were such faithful supporters of those many weekends and Annual Training periods which were necessary for unit readiness. Our intent with this work is to produce a written heritage of C Troop so that when future generations of Troopers ask: "What did you do in the National Guard Granddad?" there will be an entertaining and detailed answer.

C Troop had lot of fun as we trained hard in a tough environment with full throat willingness. This sentence requires a lot of explanation and the reason that it must be explained is so that the reader is not misled by the many entertaining stories about the training and conclude incorrectly that there was too much fun and not enough training. For C Troop it was Mission First, and to be safe.

The first reality of ACT training is that due to the sheer risk in training after dark, our training after dark was very selective. Primarily we conducted range firing operations at night. This training included Mortar Crews firing illumination rounds to light up the target area so that the tanks could engage the targets in direct fire. The Scouts and Infantrymen participated in this both by acting as forward observers and adjusting the Mortar Fire and they would rotate on an off the tank firing range or into a flank position to join the tanks in target engagement with their vehicle mounted Machine Guns. The entire Troop was involved: cooks serving meals and snacks, maintenance Section fixing whatever broke, supply section providing ammunition services, Commo Section maintaining communications between the range and range control, and within the Troop, Drivers operating their vehicles, and officers and NCOs overseeing every aspect to ensure safe and effective operations. At regular intervals the ACT would conduct a night road march involving essentially every vehicle on hand. There is a detailed account of one such night road march. Other less intense training was conducted at night, usually at Platoon level and might be limited to land navigation or conducting a raid on a competing unit. These typically involved wheeled vehicle movement to a dismount then foot movement to the objective to conduct whatever mischief had been determined desirable. There is a detailed account of one such operation.

The second reality of ACT training has to do with the overall intensity of a training weekend. I am not aware of any other unit of any type who trained as intently as did the ACT. A key reason for our training regimen was that our unit was vehicle-centric, and vehicle intensive. Most of these vehicles were full tracked armored vehicles and they were located and maintained at Camp Pickett, VA., remote from our Armory in Richmond. C Troop conducted a training assembly one weekend each month, plus 2 weeks Annual Training each summer.

The weekend was called a Multiple Unit Training Assembly (MUTA). In December and in January, we conducted what was known as a MUTA-4. The 4 is because there are 4 training periods of 4 hours each (two on Saturday and two on Sunday). A MUTA-5 would have been Friday night (4 hours) and Saturday and Sunday. We always conducted MUTA-4s in December and January at our Armory and caught up on a myriad of administrative and non-field training requirements. Nine months were MUTA-5s conducted at Camp Pickett (Friday night plus Saturday and Sunday). Finally, one month would be reserved for Annual Training (two weeks) and this would vary anywhere from April through August and the location varied usually between Camp Drum NY and Camp Pickett VA.

The MUTA-5 was a knuckle buster from start to finish. I know we all look back sometimes and wonder how in the world we were ever able to do that even one weekend, much less nine weekends a year; year after year. But we did because that was our mission and we were a team who would not let each other down; and we knew there was no other unit on the planet who could do our job like we could do our job. And besides, it was so utterly enjoyable to press the envelope of performance and get every last ounce of exhilaration out of everything that roared or clanked or went boom. Everybody, without exception, got a special buzz out of blowing things up or ricocheting projectiles off it or watching same. Allow me now to describe a MUTA-5.

The MUTA-5 could not have been successful without what we called an Advanced Party (AP), a group of soldiers who were entirely volunteers, who preceded the motor march of the main body to Pickett by about 12 hours. The AP operation deserves special praise. There was a core group who always were among the AP, from my perspective, from 1968 through 1975. Bill Talbot, John Deats, Jimmy Norwood, Whitey Johnson, L.P. Hening, Clarence Taylor, and often Macon Morris, who was a full-time employee at Pickett, were regular volunteers for this as were many others.

C Troop stabled all of their wheeled vehicles (1/4 tons, Ambulances, and 2-1/2 ton trucks) in their motor pool at the Armory in Richmond. All full tracked vehicles were stabled at Pickett and were issued to the ACT for training. Don't get the idea that all we had to do was to drive leisurely down to the Pickett motor pool like going to Enterprise rental car office and get all the vehicles we wanted. Remember, this is the US Army and there are requirements for using their stuff.

Here is the scope of the equipment which we inventoried and signed in blood for:

9 M-48 Fully Tracked Armored Tanks (further back in history the tank was the M-41)
1 each (sometimes 2) M-88 Full Tracked Tank Retriever
1 each M-577 Full Tracked armored Command Vehicle
3 each M-106 Full Tracked Armored Mortar Carrier
3 each M-113 Full Tracked Armored Personnel Carrier (further back in history the APC was the M-59)

Sounds simple enough but the task has just begun. These 17-18 behemoths were stripped of all of their On Equipment Material (OEM) which would make them capable of being destructive systems and provide lethality in war or mission training. All we had at this first step towards a successful MUTA-5 was a ride. All of them would operate and burn fuel and run over things and provide great fun but they were just a ride until the OEM was re-installed.

This would amount to several truckloads of a long list of individual items which we had to inventory, sign for, load onto our transportation, and move it to link up with the vehicles. Then we had to install it all into the tracked vehicles so that when the main body of Troopers arrived about 2200 hours on Friday night, they would fall in on a fully equipped, ready to train, piece of combat equipment. The OEM list below is not complete but is sufficiently so to appreciate the enormous task to be accomplished by such a small number of advanced party personnel, some of which would not have the skill and knowledge to install much of the OEM on a vehicle.

Most of the time the advance party would convoy the OEM to the field training site and then install the OEM on the Tracked Vehicles during the day on Friday so that we could optimize the amount of work we could accomplish during daylight hours when training began Saturday morning. The main body would then link up with the fully outfitted tanks in the field, spend Friday night with their vehicles, and arise the next morning for breakfast at 0600 and begin field training at 0700 Saturday. Typically we would train hard all day and have lunch at about 1200 and Dinner about 1800, depending upon the month and thus, sunlight.

There were a few soldiers from time to time who were in the unit for the wrong reasons and had to receive corrective attention, but these were rare and manageable. The Troop was blessed with good NCOs, creative Platoon Leaders, and willing Soldiers.

Another critical task for the AP was to draw subsistence rations to feed the Troopers a grand total of 5 full meals, water resupply, and for some missions, late night snacks. This was particularly the routine during night range firing as this mission typically went on until about midnight.

Range firing created an additional huge task for the AP in that ammunition had to be drawn from the Ammunition Supply Point (ASP). With a live fire exercise scheduled, we would draw thousands of rounds of ammunition to include Cal .30 MG, Cal .50cal Browning HMG, small arms, 90mm main tank cannon rounds, and 4.2" mortar ammunition. All of this added a separate element of danger to the MUTA-5.

Even if we were not doing live fire exercises, there was still an Ammunition Supply Point (ASP) mission to accomplish, that of drawing training ammunition which included blank rounds for every weapon below Cal.50, a copious supply of Artillery Simulators, smoke grenades, signal flares, booby trap material, and whatever else there was that would go boom and not be considered live fire. The 90mm tank cannon actually had blank ammo capability and when this device was available, we would draw a couple hundred rounds or whatever we could talk the ASP operators out of.

At the end of available training time on Sunday morning, usually about 1000, sometimes a bit later, it was clean up and turn in time. We were issued the vehicles and OEM in clean, maintained condition and it had to be returned the same way. We call this phase of the MUTA-5, TURN-IN. I do not recall ever having lost an item of OEM, although, there were several instances where we very nearly lost a Tracked Vehicle.

TURN-IN was the epitome of teamwork. The OEM was the easy part at TURN-IN because the entire vehicle crew, fully trained on every aspect of a component, was available to remove and clean each item. Once this was done, the big task, cleaning and inspecting the tracked vehicle, was on our shoulders with condition. Vehicles were ushered through what was known as the "wash point"; typically, a large concrete parking area populated with high pressure/high volume water hoses with a nozzle adjustment which, it seemed, could rip the skin off.

I don't know how many pounds of mud can cling to a tank, but it is probably measured in pickup truck loads. It is easy to imagine that water battles are simply not avoidable. Time-On-Target collaboration to focus multiple hoses on a single individual were not uncommon. They could be funny, but not necessarily to the target. Retribution would always occur, though not necessarily on the offending MUTA-5; but maybe months out. A two-to-three-hour period was usually required to have the vehicles and OEM prepared for acceptance by the issue point. All tracked vehicles cannot use the wash point concurrently and so crews rotate between wash point and having lunch and of course the wheeled convoy which will take all and everyone back to Richmond, must also be cleaned so than when we return to the Armory, the wheeled vehicles will arrive clean and ready for storage.

But it is not over yet because there is OEM on the wheeled convoy also, which must be dismounted and secured in Supply. By the time all this is done, the time is typically 1800 when final formation is called, leading to dismissal.

This, friends and families, is the outline of what your soldier accomplished on a MUTA-5, without going into the details of a training schedule. Many of the "tales" which follow, will collaterally describe in detail the nature of much of our training. We, the Troopers of C-223rd, take great pride in sharing what we accomplished. We are not aware of any past unit which has committed their stories to text with such candor and transparency. In all that we accomplished over about 7 years, very few injuries, and no deaths, ever occurred and for that, we are all blessed, and we thank God.

What did occur, during the 7 years of my association herein, as well as the 10 years or so that the ACT remained active after my departure, was the development/creation of an abiding and genuine camaraderie among the Troopers. This includes friendly competition, which is really the fertilizer for teamwork, which was an absolute necessity for our mission. The camaraderie has survived the years and is evident when we assemble annually to renew the bond of brotherhood and shared difficulty, elation, pride, and exceptionalism. We share tales of venture together, some of which naturally gets embellished at each telling. Or maybe our individual gray matter is simply improving with time.

Carroll Childers, MG, ret

C-Troop Honor Roll - July, 2019

Those who served so well, remembered and admired,
called from their duties by God

Matt Andrews
James Bailey
Ray (Rocket Man) Barlow 2012
Russ Borass 2019
Alfred (Bull) Brammer
Lindsay Bruce
Fred Bruner
Richard Burton
John Bryant
Pee Wee Daniels
John Deats
Dick Ford
Wilson Ghee
Nehemiah Givens
Les Grandis
Maurice Harver
Paul Hanson 2019
Lowell Hensgen
Mike Jarvis
Whitey Johnson
Billy Jones
Joe Kiesling
Sonny Lipscomb
Buddy Martin
Rob McGuire
Macon (Mercury) Morris
Jimmy Norwood
Dennis Pounds
Robert (Clarence) Taylor
Johnny Thompson

Basic History of the VA Army National Guard and Armored Cavalry

Carroll Childers, MG, ret

Armored Cavalry in Virginia goes back to the 1960's when the 183rd Armored Cavalry Squadron was a unit within the TOE of the 29th Infantry Division. At that time the Squadron was commanded by then LTC Ross Morris, a WWII combat veteran. LTC Morris was promoted to Brigadier General in later years. The Squadron Headquarters was in Richmond. Also located in Richmond was A Troop commanded by Captain James A. Bailey, B Troop commanded by Captain Mike Marshall, and HHT commanded by a Captain Carl Garrison. C Troop, commanded by Captain Alvin York Bandy, was located in Fredericksburg, VA.

As the US Army began to reorganize itself, it decided to retire the colors of the 29 ID and began to steadily lay the political groundwork to that end. After two attempts to retire the Division, right up to and including the full planning and rehearsal for the ceremonies inherent in such affairs, in the first quarter of calendar 1968 the Army succeeded on its third attempt. Clearly there was a lot of behind the scenes negotiation in this important event, such as all the jobs associated with the armor maintenance facility at Camp Pickett as well as the extensive training area, tank gunnery ranges which needed visiting units as customers.

The outcome of whatever the negotiations were, was that Virginia would retain one Armored Cavalry Troop. Although this Troop would be Virginia Army National Guardsmen, they would be a unit of the 223rd Armored Cavalry Squadron, 28th Infantry Division, PAARNG, headquartered in Philadelphia, PA. The VA Cavalry would be wearing the "Bloody Bucket" Shoulder Patch of the 28th ID.

In July 1968, I returned to CONUS from a tour of duty in Viet Nam, not as a soldier, but as a DOD Civilian embedded with the US Navy Brown Water in the Mekong Delta. The 29ID was retired during my time in Viet Nam. The commander of the singular Cavalry Unit in VA, C Troop, 223rd, was Captain James A. Bailey and he had asked for me by name to be a Platoon Leader in the new organization. The XO was Joe Lucas with Lieutenants Pounds, Deverell, Bugg and the other Platoon leaders. The First Sergeant was Donald Knapp, a USAF NCO, veteran of Viet Nam. Captain Bailey and Knapp had apparently been allowed to select soldiers from the entirety of the retired 183rd Troopers from HHC, A, and B Troops, i.e. the Richmond units.

Let us review the mission and TOE of an Armored Cavalry Troop. The Mission is:
- Ground reconnaissance and surveillance within, and to the front, flanks, and rear of division.
- Assistance in the movement control of divisional and non-divisional units within and through the division's area of operation.
- Assistance in division command and control by providing a positive command link between the division and subordinate elements.
- Surveillance and reconnaissance to facilitate the rear battle and economy of force missions within its capabilities.

The TOE of an Armored Cavalry Troop has changed a lot since the era of C Troop 223rd Squadron. It has changed in the type of equipment, the organizational structure, and manning levels. We were fortunate to have served when we did because the TOE of C Troop gave a tremendous capability and responsibility to the Platoon. At the Platoon level, combat capability included Armor, Infantry,

Reconnaissance, indirect fires, command and control, resupply, and maintenance.

So what did we have at Platoon Level:

Tanks (TC, driver, loader, gunner/Asst TC)	3
Infantry carrier (M59/M113)	1
Mortar carrier 4.2" Mortar	1
Scout Tactical Vehicles (M151/38A1/M114 with driver, observer)	5
Plt HQ with PL, Plt Sgt, Driver, RTO	1
Total Manpower at Platoon Level	46
Medics assigned from Troop HHC	1

    C Troop had three of these above line Platoons in addition to the HQ Command element and the HHC which provided the functions of Personnel, Supply, Maintenance, Communications, Armorer support, Mess, and Medical support.

    The Line Platoons, of course, seemed to get the most attention and were centric in terms of driving the training program, which was the metric for evaluating effectiveness and readiness. I think every Trooper recognized, however, that none of that evaluation of readiness could have been accomplished if not for the herculean support effort of the HHC.

    On the battlefield, no O-3 commanded unit is as heavily armed and as deadly and destructive as is an Armored Cavalry Troop. From the individual arms and crew operated Ma-Duce .50 Cal through the 40mm grenade launchers to the armored vehicle main gun and healthy 4.2 inch mortars, all were on highly mobile and agile motor platforms. This combination of firepower and ability to move, shoot, and communicate, enabled these warriors to regenerate the inferno of Dante on call.

C Troop demonstrated time and again, under every clime and condition God could burden us with, that we were the very best the Army had in inventory. Our unspoken moto was "We can, we will, we did that, and there is nothing we cannot do better than the rest." No challenge from competing units went without response. (We should all be in jail! But we were too smart.) We were CAV.

ONCE CAV, ALWAYS CAV. IF YOU AIN'T CAV, YOU AIN'T DONE NOTHING NOTABLE.

Alfred "Bull" Brammer
L.P. Hening

I enlisted Bull Brammer into C Troop, 223rd Cav. The Friday after he enlisted, we were in the orderly room at the Dove Street Armory getting ready to go to Fort Pickett for weekend training. Captain Childers came into the room as he was also going with the advance detachment to Pickett. (The advance group went to Pickett ahead of the main body of the Troop to draw vehicles install weaponry, radios, and all sorts of equipment needed for the training assembly.)

I introduced Captain Childers to Bull, who had recently returned from duty in Viet Nam, and Captain Childers noticed the "Big Red 1" patch on Bull's uniform. Captain Childers asked Bull..."What was your M.O.S. in Viet Nam?" Bull responded, "Pop-up target, Sir!"

Bull was SSG in the maintenance section in the 1970s under motor sergeant Maurice Harver. Steve Graves, John Mull, Matt Andrews, Hershel Carter and Wilford Portwood were also members of the maintenance section at that time.

Editor's note: Bull was a very likable guy originally from southwest Virginia. Short, round-faced and stocky, he spoke slowly with a pronounced southern drawl. Rarely seen without working on a chaw of Red Man, he was a jovial person, always smiling and joking, who made the whole unit better just by showing up.

Always Eat Last
Carroll Childers, MG Ret.

I have seen many field grade officers whose total focus in life was to become a General Officer. If they had spent half the energy in doing their assigned jobs as they did trying to impress a senior in the hopes of being recognized as GO material, perhaps they would have been recognized. Most did not. To begin with, in the National Guard there are only a limited number of GO positions in Title 32 assignments. Typically, and most often, only 16 percent of states have one Title 32 MG position and maybe 3-4 BG slots and some of the BG slots are blue suiter. Of course, the Adjutant General is a MG but not a title 32 position. (That position does not really count, as it is a political appointment in 49 states and is not directly troop-related. It is quite administrative.) Today there are only 8 Army Divisions in the NG, each having one MG slot.

My own Modus Operandi was always to be the best soldier I could be no matter what the assignment, to never ask for a position I was not assigned to and, to stay where I was assigned as long as I was allowed. I always thoroughly enjoyed the challenge of command no matter what level. As I have told people, had I not been told to move, I would have retired as an Armored Cavalry C Troop commander (0-3). I stayed there 49 Months as it was until finally the Brigade Commander came for a visit and told me he was moving me to his staff. That is when I converted from Armor to Infantry.

So I worked diligently, led from the front, created challenging training programs, and demanded performance at all levels. I endeavored to set the example, shared all pain with the troops, and recognized them at every occasion where they deserved it; no false praise.

As some other leaders may have, (I never conducted a poll) I always ate last, no matter what else was going on or where I needed to go. And if the line looked too long so that I would not be able to make a meeting or linkup, I would just fore-go that meal.

I remember on one occasion it was the evening meal at Ft Pickett. I knew that both 2nd and 3rd platoons had tanks mired to the fender wells and tracks broken and that the M88 was out there trying to recover the tanks. Standing off to the side of the chow line were the Platoon Leaders of these two platoons, talking and spitting tobacco and probably lying about something. I walked over, knowing the situations of their units, and asked them how the recovery operation was going.

One of them boldly spoke up to proclaim "Sir, the Motor Sergeant says he will have the job done by 1900 for sure."

My reply was "So what are you doing here in the chow line when your tank crews are still in the field? Go tell the Mess Steward how many people will be reporting for late chow and you get *%^$(* back to the recovery operation. "Oh, and I will be eating with you."

And off they went in a cloud of dust in two M38 Jeeps. The two LTs learned a lesson that day. They still talk to me; and that was almost 50 years ago. And the Mess Steward, got such a big kick out of his part in that lesson and still reminds me of that story now and then. He makes an excellent blackberry wine and always brings me a few bottles. Such is the nature of the Guard. It too has a strong camaraderie.

And Then There Was The Time When...
SSG Anthony Burke

We were at Camp Drum to conduct Annual Training (AT). We were doing tactical maneuvers and pulled into a defensive position for the night. It was already dark by the time we pulled in, and it was one of those moonless nights. We set up a circle, with each platoon protecting a sector, and congregated inside the circle. My M113 APC, like the other track vehicles, was facing out from the circle. All night long, my crew and I always exited the APC from the rear, and not the front. (It was common for the driver & TC to slide down the front ramp, because it was easier than crawling through and out the back.) The next morning when there was enough light to see, we realized that the vehicles were parked not more than 2-3 feet from the edge of a quarry. Had we positioned the vehicles any further in the dark, or if any of us had gotten down the front of the vehicle during the night, the vehicle and/or personnel would have gone over the edge.

We were conducting AT at Camp Pickett. By coincidence, so was the artillery unit. At the end of training one day, our unit was in the motor pool securing the vehicles for the evening. The artillery unit was on the firing range that was closest to the cantonment area; and the motor pool was in between the two. As we were unloading gear from the vehicles, we heard the distant report of an artillery round being fired, immediately followed by the roar of the round sailing right over our heads and exploding somewhere on post in the direction of the cantonment area. Shocking as that was, the first report was followed by a second one and a second round going over us and exploding on post. At this point we all "un-assed" our vehicles and "dee dee mau'd" out of the area most "ricky ticky." Later, we came to learn that one of our units had received a message to call his boss, and had gone to the cantonment area to use the only public phone next to the NCO Club. The Club was located on one corner of a vacant lot on base, with buildings surrounding the lot. As he was talking on the phone in the booth, the first round hit in the vacant lot, soon followed by the second one. By the grace of God no one was injured or killed, and no buildings were damaged, especially the NCO Club.

We were at another AT at Camp Pickett and Jack Page, tank section leader, 1st Plt, was up for re-enlistment, but allowed as to how the only way he'd re-up was if the Brigade Commander came to Pickett and personally swore him in. Well, some calls were made and, sure enough, one day at formation, the Brigade Commander was there to swear in Jack.

Whenever the Troop would motor march to Camp Pickett for a MUTA5 weekend, we would leave from the Dove St. Armory on Friday night and head down state route 360 to the Blackstone exit. Along the way was a truck stop diner called the "360 Truck Stop", where the convoy would stop for a break. It was a welcomed opportunity to grab some extra snacks/ drinks for the weekend, or even a quick meal.

We were headed to Camp Pickett for a MUTA 5 weekend training exercise. As usual, we left from the Dove St Armory at night. It being the Oct-Nov season, it was already dark. Also, as usual, we convoyed in the unit's wheeled vehicles (quarter tons, 2 1/2 tons, fuel truck, 3/4 tons, etc.) and they had just been repainted. We motor marched down Rte. 360 and turned onto the other smaller state route 153 headed to Blackstone and Fort Pickett. This happened to be my and some other guys first training weekend since returning from active duty. I and one other newbie caught a ride with Ron Portwood in the fuel truck. This secondary road was bumpy and curvy and rainy; so, the truck was bouncing and lurching all over the road, and Portwood was smoking a cigarette the whole time. We thought we were going to die on our first drill.   The night continued. About half way down this road, we notice the other vehicles in the convoy on the side of the road, broken down. As we passed, we commented on the sight, and Portwood said something like, even though they were broken, they sure were pretty in their new paint.

This same night continued to prove notable. When we finally arrived at our bivouac site, all our duffle bags, which had been put into one 2 1/2 ton truck, were dumped out into a large mud puddle. Meanwhile, as we spent the night in the rain, Cpt Bailey went into Blackstone to spend the night in a hotel. The next day Cpt Bailey complained that the rain was so hard that he had to shut the window to his room. We were not amused.

On another AT, the scout sections were doing scouting maneuvers in an area that was hilly and thickly wooded terrain. Our scout section was moving slowly up a steep grade as we were "breaking bush" at an interval of about 20ft, when suddenly, the M113 in front of mine just disappeared. Quickly moving up the hill to find out what happened, we were shocked to realize that we were at the crest of a razor back ridge and the M113's rear was sticking straight up in the air just over the crest. A tree had stopped it from going any further, saving the crew's lives and the vehicle. We hooked up tow cables and pulled it back over the crest. Another close one.

Then there was the time when we were doing night tactical exercise with night vision devices on the track vehicles. This has been previously reported on, but a footnote is helpful. We were given the afternoon to set up/ install/ test the vision devices in preparation for the night exercise. Well, we had never worked with these devices before, and did not realize that there was no way to confirm they were operating correctly during the daylight hours (since they were NIGHT vision devices). Duh! All we knew was that they were installed. So, when it came time to move out that night, they didn't work and we couldn't see a thing. It was a miracle there were no other "exciting" incidents besides the tank almost going off the bridge.

Then there was a firing evolution at Camp Pickett, doing tank main gun fire exercise – both day and night. That day the scouts were practicing calling in night illumination to support the tankers. There was one humongous supply of 90mm rounds stacked on the ammo tables. The day's exercise was to zero the main guns and lock in the coordinates where the 55gal drums full of fuel were to be placed for night fire. This all went fine. Then there was night fire. As usual, no moon. As the evening fire progressed, it became more and more breezy, and then windy.

There was no moon, because, unknown to us, a storm front had moved in after dark; and one hellacious lightning storm unleashed right on top of our position. Lightning was striking all around us and the ammo. We quickly ceased fire, closed the range, secured vehicles and "ran" in full retreat back to the bivouac site, while more lightning smacked the ground all around our vehicles as we dashed back to camp in the blinding rain. Well, our bivouac site was in a copse of trees next to a big field. Lightening was coming down all around the trees/ camp site. I have never seen so many people fit into vehicles as that night. The M88 tank retriever had at least 10 people inside it. One trooper, lying prone in his pup tent, was lifted off the ground by a lightning strike. When we got up the next morning, there were large tree limbs on the ground that had missed some of the tents by mere feet. So, whoever said drills and AT were routine?

When the Troop had a locker room in the Dove St. Armory, each person could keep his gear, instead of having to carry it back and forth between home & drill. That was convenient. Additionally, one locker was reserved for storing snacks, drinks & other "refreshments" that were sold to troop members. This was the "club" locker; and on weekend drills at Pickett, the "supplies" were packed up to go with us. During the weekend daytime activities, one ¾ ton would be assigned to as the "gut truck" so it could follow the troop through the maneuvers. Then at night, the "club" would be set up in one of the barracks. While not exactly army issue, this added feature certainly made the weekend training much more enjoyable, and needed, for washing down all that dust from the day's endeavors. It added to the "esprit de corps" in general.

The club truck was also a welcomed sight in the field, especially because the cooks could not read a map; and therefore, when all three platoons were in different training locations, the cooks couldn't find the locations in order to deliver the meals. It got so bad that, at meal time, each platoon would send a scout team back to the barracks to lead the cooks to the locations. Until the meals could arrive, the club truck helped fill the hunger gap.

Annual Training, Camp Pickett 1973
SPC-5 Mike Thomas

Tank Range Five

As the CO's driver, I got to witness the full-range of C-Troop training. Watching our complement of nine tanks lined up and firing on Tank Range Five was an awesome experience for a newspaper employee accustomed to working in an office all day. I'll never forget seeing those heavy tanks "rock" in recoil from their main gun blast. The crews fired "HE" rounds at old vehicles used as targets scattered around on the impact area hillside thousands of yards down range.

At (AT) Annual Training, one full day and night was devoted to direct fire exercises with the main, 90mm tank guns. Tank Range Five at Pickett was perfect for this training as it provided great target views.

Mortar Support

While the tankers were busy burning up rounds in daylight, our three mortar carriers were set up in battery formation at another firing point further down Pickett's main tank trail. This firing point was situated at a right angle to Tank Range Five and allowed our mortar crews to see the same targets the tankers were firing at from Range Five.

The mortar guys and our scout sections spent most of the day working on fire missions of their own. After initial set-up and rough orientation of the mortars, the plotting board was made ready awaiting fire-mission instructions. Cary Hatcher and Rob McGuire were among those who expertly handled the "math" and issued firing instructions to the assembled mortar carriers. Our scouts, led by Roddy Davoud and other platoon scout section leaders, from their forward OP's, practiced calling in fire missions on targets such as "troops in the open" or "vehicles in the open" and gave target grid coordinates. Then, spotting impacts, the scouts radioed back fire adjustments, such as "add 100 (yards), right 100" and continued to adjust fire until the target was engaged. Then, the command…"fire for effect!"

Night Firing

Preparing for night firing began in the late afternoon with tankers unpacking and loading the rest of their rounds. At the same time, over at the mortar firing point, test firing of parachute flare rounds began. These special rounds carried a long-burning flare that would be released from the host mortar shell in mid-air. Then, the three flares, one fired from each tube, floated slowly to the ground suspended by its parachute. The parachute's hang time and flare's burn time was set by the mortar crews with the objective being for each flare to light high and burn out before it hit ground, just as another flare would light to replace it high above the target area.

Like controlling a continuous fireworks display, using a stop watch to track the flight time to shell burst and flare burn/decent time, the mortar crews would "hang" and then "drop" consecutive volleys. Almost without interruption, the tankers' targets were constantly illuminated during the hours-long exercise; an impressive demonstration of our mortar guys' expertise!

Flash Fire

With the flare test firing complete, and the weapons' settings dialed in, the crews unpacked the remaining flare rounds — about 40 rounds per track — stacked them two-high in rows beside each carrier. The rounds were further made ready by removing some of the "cheese patches" and shell extensions so that they all had the precise quantity of propellant charges, which together with the angle of elevation of the mortar tube, would determine the three dimensional location over the target area where the flare would ignite. (The cheese patches came in 2X2 inch bundles. Their yellowish tint and waxy appearance made them resemble a package of cheese slices.) Excess "cheese" elements, which were made of highly flammable, phosphorous-infused material, were placed in a burn pit to be disposed of the next day.

After all the flare rounds were ready, a tarp was draped over the nose-end of the stack of shells and the crews moved back to a shady spot, at the rear of the firing point, to wait for the night-firing exercise to begin.

Just before dark, without warning or any obvious cause, the "cheese patches" on one of the shells ignited, setting off an instant chain reaction along the all the other shells in that line...a super-quick, flash fire! The white flame went up at least twenty feet and that sight, plus the accompanying roaring, sizzling sound, sent the dozen Troopers scrambling. The flash fire lasted only seconds but everyone thought at that moment that we had had it! The fire would explode the entire mound of flare rounds and wipe us all out. But, thank goodness, the stack of flare rounds did not go off and no one was hurt.

This incident was investigated by post authorities but a definitive conclusion was never reached as to what caused the ignition and flash fire.

AWOL Trooper: December Drill, Dove Street Armory

C Troopers rarely spent training time at the armory. The December drill assembly, however, was usually held on Dove Street. This drill period was spent on weapon and equipment maintenance, classroom instruction, close order drill, etc. Marching in the confines of the armory led to some interesting cadence calls from the NCOs taking turns to direct their platoons. Normal marching orders called out were column left, march, column right, march, platoon halt, and so forth. Due to the relatively small drill floor, the cadence caller had to be quick thinking to avoid mayhem, giving rise to the occasional necessary order of …"Come away from that wall, March!"

Opening formations began with roll call and on one particular December Saturday the First Sergeant called one Trooper's (trooper Doe) name several times with no reply. After formation, the First Sergeant called aside two of the missing soldier's buddies to ask if they knew why Trooper Doe wasn't present.

"Sarge, he's around…in fact, we just saw him downtown an hour ago."

The First Sergeant said, "I want you two to go snag him and bring him to this armory! He's in serious trouble. Wait right here for a minute."

He walked down the hall to the MP unit's orderly room and came back with two MP helmet liners and two pairs of MP shoulder badging.

"Here, put this stuff on, get in a jeep and go get Trooper Doe and bring him back here!"

They did just that…

Trooper Doe was taken to the Captain who lectured him about being where he was supposed to be. "No excuses. You can't just decide to not come to a drill." He reduced his rank by one grade and then dismissed trooper Doe to report to his platoon sergeant.

A bit unconventional, but Trooper Doe was never absent from a drill muster again turned into an outstanding soldier.

Bailey-isms
LT Bryce Bugg

Ok, here's the deal.....Spring 1970, we're having an O meeting prior to AT. I'm brand new to the troop. I didn't know one kinda jeep from another. Bailey is prattling on about how great his old jeep (M38?) is, while the others are sure his jeep will break down. We all had the new Jeeps built by Ford. Bailey bets each of us $20 that it won't break down. He was on the side of the road before we got to Ashland.

We get to Ft. Drum and Bailey goes out and spends about $100.00 on food and booze so that we could entertain some Grand High Kahunna (who knows). He decides to pass the hat amongst us. Now, knowing full well that getting $20 (each) outta Bailey was a dream at best, I announced that I was square, referring abck to the bet surrounding the jeep that had never been remitted. The other 3 took about a split second to realize that Bailey had well and truly screwed himself. Sweet!

When the film came to Richmond, the 5 of us (see above) got together and went to see Patton. Bailey was never the same afterwards. I don't remember the pistol, but I do remember how the leather gloves and the riding crop showed up at the next drill. It's probably a good thing that George C. Scott didn't appear wearing a Tutu........you get the picture.

Another time we were at Pickett camped in the woods. About dark, Bailey disappears. That night it rained like a cow peeing on a flat rock. Bailey shows up for breakfast. We were discussing the rain and he chimes in with,

"Rain? You talk about rain. It was raining so hard over at the 2/111 BOQ that I had to get up and close the window!" Nothing like leadership from the rear.

The call signs for the leadership: 6, 5, 1-6, 2-6, and 3-6 were actually shortened from the complete call sign A complete call sign for C Troop would have been Raspy Squabs 6, meaning the C Troop commander.

Blow the Field Latrine
Carroll Childers, MG, ret

I was commissioned a Lieutenant of Cavalry upon graduation from OCS Class # 6 (August 1964), Virginia National Guard. Not horse cavalry, but armored cavalry. There was a saying among Cavalrymen; "If you ain't Cav, you ain't ****." A meaningless statement if you examine it, but the intent is that the Cavalry is the best of all Military Occupational Specialities (MOSs); it is braver, more innovative, more daring, more durable, greater risk takers, capable of missing meals, able to continue a mission endlessly, and always accomplish it. Nothing was too tough and to prove it, since we did not have a war to fight, we sought challenges to overcome, dirty tricks to play, demonstrations to perform, senior commands to outsmart, and above all, to be able to outdrink any other group at any celebration or affair that we were invited to or not invited to.

I served first in C Troop 183rd Armored Cavalry Squadron in Fredericksburg, VA. In 1968 while I was in Viet Nam, the 29th Infantry Division was retired and when I returned from Viet Nam I had an assignment in C Troop 223rd Armored Cavalry, associated with the 28th Infantry Division. The 28th was a Pennsylvania National Guard Division and the Squadron Headquarters was in Philadelphia. Only a single Troop was in Virginia, C Troop, 1- 223rd in Richmond. We trained during annual training on a rotating basis between Ft AP Hill, FT Pickett, and Ft

Drum NY. Eventually Ft AP Hill was abandoned as a training site for C Troop as all tracked vehicles were relocated.

The C Troop commander (TC) was one Captain James A. Bailey. Jim was a native Virginian who had spent several years on active duty then left the service and came to the Virginia National Guard and took over as a C Troop commander. When I returned from Vietnam, Jim had held a Platoon Leader slot open to offer to me. He had seen me operate in the 183rd and had the kind of qualities that he wanted to surround himself with. In hindsight, he wanted people who could make him look good. His complete story is another story but for now, the subject is blowing the Field Latrine. Jim was always looking for something really colorful to do and by his example, he was encouraging his junior officers to do the same. So Captain Bailey influenced me a lot in being somewhat of an exhibitionist and I admit it.

I became the TC within a few years and took the Troop to Ft Drum NY (still Camp Drum NY at the time) for annual training. In the field, my C Troop Command Post (CP) was located such that our route out to the main tank trail and to the road network giving access to the remainder of the post, went past the Squadron Headquarters CP. We took some pleasure in creating dust storms that often drifted into Squadron CP. They complained; we promised to drive slower; multiple times.

The Squadron Command Sergeant Major was quite proud of the toilet facility that he had installed for the Squadron CP. It consisted of a series of five 55 gallon drums buried in a trench. Over these barrels had been constructed a wooden bench that accommodated five soldiers who could sit to eliminate their bowels. A hinged lid opened to reveal a hole in the bench that was provided for each position. Around 3 sides and top was a canvas tent and across the front was a double mosquito net. This net provided ventilation and because it was doubled, provided a surprising level of privacy from outside viewers. It was almost elegant……..and therefore an immediate target.

Strike preparations began almost immediately upon discovering this facility. We talked about what and when we might do something to the "palace latrine" over the first week, primarily while drinking beer after the duty day. Oh, yes, the Cavalry did not do much training at night because of the risk of injuring someone or damaging equipment. We did not have night vision of any kind so any sort of complex training was rare indeed.

The decision was finally made to blow the latrine using "artillery simulators." These are truly ingenious devices and there seemed to be an endless supply. Though intended to do exactly what the name implies, to simulate incoming artillery with a large explosion and puff of smoke, we found uses too numerous to discuss in this piece.

We had only a few rigid rules for the employment of artillery simulators;
- Don't hurt anyone, including one's self.
- Don't throw one inside a vehicle
- Don't use with any form of accelerant (gasoline, etc)

I think it was Saturday night, the middle weekend dividing the two week annual training period at Camp Drum. Three of my Platoon Leaders and I were returning to our CP from a beer run in my command jeep. Of course we had a bag of Simulators under the seat and of course we had sampled the beer before buying it. We had a trailer attached to the jeep with a load of beer for the troops so what a perfect time to blow the much vaunted headquarters latrine.

We drove slowly past the Squadron CP. It was dark; perhaps 2200 in the evening. Coasting to a stop so that there was no possibility of a squeaky brake, we divided up simulators and I gave assignments. I would drop 2 in the right two holes; LT Pounds would drop 2 in the next two holes; Lt Deverell would drop one in the final hole. Then we would move expeditiously back to the jeep where LT Bugg (the junior LT) was waiting behind the wheel to effect the getaway. Simulators have a 10 second delay. Plenty of time.

We made our way to the latrine in the moonlight and checked it out. No one was occupying it. No one was approaching it. We all stood together in our positions and simultaneously pulled the strings to ignite the simulators, then dropped the individual simulators in the holes. We made an exit. I sent the two LTs to the waiting jeep and I stayed behind offset from the site to ensure that no one showed up to use it so I got the best view of the final result. It was such an explosion that I was briefly captured by the uncalculated beauty of it; we had not counted on the methane gas involved. The LTs back at the jeep were exclaiming in grandiose terms of unbelief, and perhaps fear. I ran the 50yd dash and told Bugg to "get the hell out of here." Maybe we had gone too far this time?

We bombers arrived back at my CP and went to my tent to have a beer and marvel and the genius and stupidity of the strike………and, how the hell we were going to avoid courts martial for it. Denial! No one could actually prove who did it. Only the four of us knew. Rumors of trouble began to creep into my CP about dawn the next morning. My First Sergeant was quietly furious. "I know you did it sir. It's got your trademark on it. I'm gonna see if I can fix it."

Ah, the value of a good senior enlisted man.

The First Sergeant took a work party over to offer to fix the damage based not on guilt but upon willingness to assist to fix the prank of an unknown party. By the time the Squadron Commander and staff had returned in the evening of Sunday from leave, all was back to normal and the whole event was just another good story with the perpetrators still unproven. The event was somewhat of a report of caution; perhaps a warning that such an event would not be tolerated in the future; but behind all of the official rhetoric was some form of Squadron pride that someone had blown up their super shitter. Wish I could remember the name of the Squadron Commander. LTC Cliff (last name was I am pretty sure: Boyer.)?

He was a Postmaster in Philly and a super Squadron Commander by every measure. Fortunately for C Troop, there was another Troop (A) in the Squadron that presented LTC Boyer far more conflicts to resolve than did we. They are an entirely different story and I am not sure I will attempt to cover them in any detail but there is a published book describing them in far more politically correct terms than I could ever convey and the title of the book is The Gentlemen From Gloucester.

Camaraderie
Spec-4 Alan Hughes

Alan Hughes drove Johnny Thompson's tank but his civilian job was as an engineer working at for the Naval Weapons Testing facility at Dahlgren, Virginia. In addition to his tank driving and engineering skills, Alan was also fantastic bowler, good enough to consider trying the PBA tour at one point. Alan's other talents include golf and poker.

As soon as the duty day was over, a bunch of regular card players would seek each other out and ante up. Sometimes the game would be in the Club area or a barracks. The favorite spot of all was the mess hall and they'd head over there ...if SFC Buddy Martin wasn't watching.

Betting consisted of nickels, dimes and quarters, three raise maximum. The stakes were set to keep the game interesting without breaking anyone's budget. Someone might win or lose $25 in a session.

We played a variety of standard games: five card stud and draw poker, Chicago plus a bunch of variations like "Flaming Cross" and Alan's favorite game, "Mississippi Mud." Mud was a split pot game, best high hand and best low hand divided the pot. No one knew where the game got its name, only that that's what Alan dubbed it. (Fifty years later, we still play a hand of Mud now and then in Alan's honor.)

At summer camp at Pickett, in July, 1971, a new player sat in the game. He casually mentioned that it was his wedding anniversary that day and that his wife, eight months pregnant, was staying with her parents in Petersburg while he was at AT. While most troopers had their POV's at Pickett, this fellow had driven a Guard vehicle in convoy from the armory so he had no way to go see his wife. Alan, who had just met Trooper, without hesitation, reached in his pocket and slid over his car keys ..."Here, it's the red Mustang...go see your bride."

C Troop Call Ups

Hurricanes
SPC-5 Mike Thomas

Hurricane Camille arrived in Virginia on August 19, 1969, one of only three category five storms ever to make landfall in the United States. One of the worst natural disasters in Virginia history, the storm produced what meteorologists at the time guess might be the most rainfall "theoretically possible." As it swept through Virginia overnight, it seemed to catch authorities by surprise. It served as a lesson that inland flooding could be as great a danger as coastal flooding during a hurricane.

Virginia rivers quickly overflowed their banks bringing havoc all over the state. The James River flood spread water over downtown streets in Richmond. Below MCV and throughout the bottom was totally underwater. Oil and gasoline storage tanks near the river were surrounded by the flood water. Richmond's sewage treatment plant was inundated. Water supplies and communications were compromised as were many transportation routes.

The governor called up Virginia Army National Guard units including C-Troop. The Dove Street armory became our operations base for the next four days. The drill/assembly bay became a huge barracks for troopers with cots set up from end to end. Our cooks set up their field stoves out back and cooked three meals a day. Trucks and jeeps were brought in from Sandston to move troops back and forth from their assignments. C-Troopers assisted Richmond policemen in public safety missions such as traffic control, acting as guards at closed off streets and to provide a deterrent to looters, especially in the downtown area and residential neighborhoods near the river. Troopers also helped business owners move their inventories to higher ground.

Half a dozen C-Troop recruits, who'd not been to basic training, were called up during Camille and were sworn into the service by an officer. Having no training or even uniforms, the recruits were assigned to the orderly room, supply room and mess areas. Hurricane Camille provided these guys a quick understanding of what being a member of the VANG was all about. Next time, the recruits would be more involved...Hurricane Agnes saw to that.

Hurricane Agnes came calling in June, 1972. It hit western Florida as a relatively weak Category 1 hurricane on June 19, 1972, then swirled off to Virginia on June 21 as a tropical storm. The water was far worse than any wind. The storm dumped 5 to 14 inches of rain in the western half of the state which caused severe flooding in the cities along the Potomac, Rappahannock, Appomattox and James Rivers. There was no flood wall in place in those days so, again, low lying areas long the James River, including Shockoe and Fulton were inundated.

C Troopers were again assigned to support local authorities in traffic control and looting prevention roles. Some C-Troopers were sent to the city jail property to fill sandbags, working alongside inmates in an impressive assembly line operation at "sand-bag central."
Troopers helped business owners save the contents of their stores and warehouses by loading up goods that were in jeopardy.

Fulton Bottom was particularly hard hit by flooding. Residents were told to evacuate the area and many did, but pockets of residents stayed in place, afraid that their homes would be looted if they left. C-Troopers were placed on sentry duty along the perimeter of the neighborhood, one Trooper posted every city block or so. Guards were doubled after dark. At first, Troopers were armed with M-16's but had no ammunition. Apparently the idea was that looters would be discouraged seeing a soldier with an M-16...no bullets necessary. That plan lasted about a day, and was replaced by a great compromise. Local political leaders didn't want Guardsman with automatic weapons and full magazines on city streets. Guard leaders argued that Troopers needed to be able to defend themselves if necessary. So, on day two, each sentry was allotted one round of ammo but the bullet was to be kept in his fatigue shirt pocket, not loaded in the weapon.

This came to be known as the "Barney Fife Compromise."

(Editor's note: Another Trooper had a slightly different take on this ammo-rationing story. Instead of one round of ammunition that was to be kept in the soldier's shirt pocket, his recollection was that each sentry was issued three rounds, loaded in a magazine that was taped to the upper left side of the suspender strap and not to be removed (un-taped) from that position unless a life or death situation occurred. This Trooper's recall is engrained because of getting braced by the XO for cutting the tape and locking and loading the magazine. Seems the Trooper did so while acting as backup for a city policeman entering a looting suspect's abode.)

The Big Snow
Carroll Childers MG (USA Ret.)

I will not even hazard a guess as to what year a massive snow storm and the creation of unpredictable drifts in certain locations took place. One thing I know for sure is that I had transferred from C Troop 183rd Cavalry in Fredericksburg to C Troop 223rd Cavalry headquartered at Dove Street Armory, Richmond. I am pretty sure I was the C Troop commander as opposed to having been the XO to Jim Bailey. I also am able to recall several specific soldiers who were called up for this duty. L.P. Hening, Ducky

Sullivan, Frank Brown, Steve Graves, Carrolll Childers, and perhaps another 15 or 20.

The duty was indeed unique as was the relatively small number of soldiers who were called to duty. The head count was probably about one Platoon in strength but they were all specifically selected because of their Military Occupational Speciality, their dependence to operate in small teams, and to understand the age old value of communicating, operating tracked vehicles in unique circumstances, and the ability to endure long hours under stress.

It was one of those once in a lifetime storms which came out of nowhere, centering on Loudon County VA. Expectations were unpredictable so the call out force were largely more senior soldiers. The other odd factor was that those called out rarely operated together, rather, some small number of soldiers (typically 2-3) were issued an Armored Personnel Carrier of one vintage or the other, and dispatched to a central civil authority control facility to assist them as required. We had a mix of APCs to include both the M-59 and the M-113. Unfortunately I and my team mate (LT Frank Brown) drew an M-59; certainly not my choice. The M-113 was a far more capable machine for the potential tasks at hand.

I no longer recall who was in charge of the whole affair in Richmond but overall it went well and whenever some poor citizen snowbound up a dead end valley and running out of whatever supplies people run out of, somehow the loop was always closed to get a response team en route in a full tracked APC with the right supplies.

Much of the response effort had a connection to electrical power, or the lack thereof. I recall specifically that a group of call ups including Ducky Sullivan, LP Hening, Steve

Graves and a few more I can't now recall, were associated with transporting linemen into power outage points to make repairs. This entire team were the guests of a large hotel complex at the relatively new Dulles Airport. If anybody tells the tale that the Officers had the best deal in this response, I have a different experience.

Overall it was about a week of fun and games in the snow and learning new methods to remove an APC from an unplanned snowbank. Fortunately we had both vehicular mounted radios as well as a few perhaps PRC-77's or equivalent and these got us out of a lot of troubles.

Probably the most exciting response that Frank and I conducted was to "rescue" an un-named landed gentry who held the Governor's phone number in his rolodex and called in to report the onset of some serious medical issue that worried him as he absolutely could not get from his most amazing quarters to a snow-cleared route to assistance. We received a radio call to report to a local Sheriff office to be briefed on the situation. Fortunately the Sheriff had a US Coast and Geodetic Survey map on the wall under plastic and had circled the location of the civilian in distress. It was about midnight as we discussed a means to make an extract. The Governor had already authorized an Air Evac at first light so our job was to figure out how to make out to the residence of the gentleman in distress, coordinate with the Air at first light, and get the casualty to the aircraft.

Fortunately some of the Sheriff's Department were thoroughly familiar with the target residence and the terrain in the area. The residence was essentially in the middle of a large tract of land and there was an associated long relatively constant altitude ridge so that by the nature of the storm and winds, a ridge approach to the residence promised the amount of snow depth that the old M59 might tolerate. So off we went, Frank and I and a bunch of comfort items for the distressed patient.

About a mile from the residence we encountered one of those areas that a terrain map often does not account for. It was still dark and overcast so we decided to take a few minor items and the PRC77 and hike the remainder of the way to a residence that we were only guessing would be at the end of the trail that the 59 was now embedded in. When the residence first came into view, lots of lights on all around, I remarked to Frank "Is that a house or a hotel." A few minutes after I rang the doorbell, a butler in full Hollywood dress slowly opens the door and inquires how he might assist us.

"Frank, I am not sure I want to talk to this guy. I outrank you, so you 'splain to him who we are. Then tell him I want a cup of very hot coffee"

The butler did say, "Excuse me please" before he closed the door and I heard the discrete little mechanical click indicating Frank and I are still outside admiring the marble columns. Turns out though that locking us out was only the beginning of the insult. Maybe 10 minutes passed and the now familiar mechanical click again gets my attention.

It is the Butler again only this time he has a platted leather leash and on the other end is a dog. Frank and I look at each other; he points at me and I point at him. Then I point to my wooly cap with the Captain Bars. Frank takes the dog down the slate deck between the Marble Columns and the brick walls. I stuff my boot into the door closure as the Butler is attempting to deadlock me out again. "If the master of the estate is up to the task, I would like to share with him the plan for his extraction by helicopter, probably within the next hour."

"Yes sir, I will share that with Mr. whatever his name was"

Sure enough, within the hour my PRC77 came to life and asking me to respond when I can hear his Whop Whop. Soon I heard him, then saw him, gave him an azimuth to him, and told him I would be popping purple Smoke for touchdown. Bird down, a surgeon stepped off and followed me to the front door. Frank had the door open and drinking my coffee.

After a quick visit with the patient the Surgeon asked me if the crew could give us a ride someplace. No sir, we have a stuck M59 that needs our attention.

C Troop Memories
Neil Henning

I enlisted in C Troop, 1st Sqdn, 223d Cavalry in May of 1971. What follows is a few memories of the ten years I spent in Virginia's only Cav Troop.

No doubt a group of anecdotes will be provided by other troopers, I thought it might be interesting to offer some of my own.

On one occasion Johnny Thompson decided to maneuver his tank across a steep stream bed instead of using the bridge. He threw a track on a large downed rotten tree and the tank came to a standstill. Then all of a sudden there was much hollering and swatting as the tank had opened a large yellow jacket nest in the tree. Thompson = 0, yellow jackets = 1

For our annual MUTA 5 weekend which was our annual "tactical" weekend before we went to AT rifles, pistols and machine guns had to be transported to Ft. Pickett. One year there was no room on any military vehicles and they had to be transported by POV. I left the armory with the trunk of my car loaded to the gills with all manner of firearms for the trip to Ft. Pickett. My platoon sergeant told me to drive under the speed limit, do not attract any attention to my car and do not stop until I reached post.

The club truck was a welcome sight in the field especially since our cooks could not read a map. During training all of the platoons were in different locations so the mess team had to feed in three different locations. However they could never find us. It got so bad that each platoon had to send a scout team back to the bivouac area to guide them to the platoon's location. Until they got to us with our meal, the club truck helped fill the hunger gap.

Every year the troop held a Christmas party on the Saturday night of the December drill. It was a nice affair as we had an evening meal made up of the army luncheon menu and augmented by food brought in by the troopers and their families. This could include venison and other wild game from the lauders of troop members. It was usually well attended but one year saw a large surge in attendance. SSG Whitey Johnson, though married, was quite a ladies man. He announced that for that year's party he was actually going to bring his wife. Since no one had ever seen her at any troop function before, a large number turned out just to see what she looked like.

Nicknames ~ the following is a list of nicknames for some of members of the troop. Last names are not included, they are in alphabetical order, see if you can remember who was who:
- Bull
- Big Duke
- Deer Slayer
- Dip

- Ducky
- Dynamic Denise
- Greasy Grass
- Hollywood
- Mercury
- Moody
- Moon Dog
- Old Man
- Rocket Man
- Tin Indian
- Toy Truck
- Stumpy
- Woodsy

Personal tool kit (Items you needed but had to purchase because they were not supplied by the Army):
- Hammer
- Nails
- ¾" open ended wrench
- Screw drivers (flat head & Philip's head)
- Chain
- Channel locks
- Pliers
- Needle nose pliers
- Duct tape
- Wire cutters

The shame of this is that come to find out every wheel or track vehicle was authorized a tool bag which was kept back in supply. Why weren't the tool bags given to the vehicle commander? I was told the officers were worried that we might lose or break them. Therefore when it came to necessary tools, you were on your own.

Personal equipment (Items you needed but had to purchase because they were not supplied by the Army):
- Machete
- Compass
- Flashlight
- Candle stubs
- Goggles
- Extra first aid pouch
- Sheath knife
- Poncho liner
- Bandannas
- OD T shirts
- Plastic cigarette box
- Map case
- Grease pencils

I am thinking that Neil lost a lot of stuff so it just seemed like he had to provide his own.

Every vehicle had a cooler provided by the vehicle crew. They were painted in camo colors to blend in. There was an unwritten rule about no one other that vehicle crew members ever opened or looked in your vehicle's cooler. This was always respected.

Theme song (as played on WGOE radio in Richmond) = "Rose of Cimmeron" by Poco

"Training Films" in the old Cavalry Troop
Trooper Bob Bruce

When I enlisted at 17 years old (with parents' permission) in Bravo Troop, 183rd Cavalry Squadron in 1964, I was a senior at George Wythe High School in South Richmond and an NCO in the school's Cadet Corps

My first official meeting on a Thursday evening at the old Richmond Light Infantry Blues Armory was a piece of cake. Our commander, the legendary Captain Mike Marshall, made sure I was put into a good squad with a kindly NCO as mentor. This was none other than George Norseffe, a real character, the oldest buck sergeant in the Squadron and maybe in the whole Virginia Army National Guard.

SGT Norseffe's job wasn't all that hard in my case because I showed up in a complete and correct set of starched fatigues and spit-shined boots, was well versed in military customs, courtesies and close order drill, and barely smart enough to keep my mouth shut. So far, so good.

At the end of the evening drill, I was graciously invited to the "RLIB Club" and this thirsty new Trooper was pleased to find that a cold bottle of Coke cost only a thin dime. Oh, and an equally cold can of beer was a just a quarter and nobody was checking IDs.

Even more interesting was the big, military-issue 16mm movie projector, set up in one end of the club with a reel of film ready to run. After a few minutes most everyone had their beverage of choice and settled down facing the movie screen. Figuring it was the proper thing to do, I took a chair too.

Well, to say the evening's "training film" in black and white with sound was nothing like I had ever seen or heard in all my 17 years is an understatement. Without going into any unnecessary details, I'll just note that I learned a lot about the opposite sex in a few short but vividly memorable minutes.

Now, men (and boys) being what God and the military made them, the informal "training films" tradition continued on weekend drills and "summer camp" for the entire six years of my initial enlistment. First in B-183rd Cav and then after re-designation as C-223rd Cav.

Real Cavalry Movies

Now, being a big fan of real Hollywood war movies, it occurred to me that great Horse Cavalry classics like Errol Flynn's THEY DIED WITH THEIR BOOTS ON and Henry Fonda's SHE WORE A YELLOW RIBBON, would be a bit more wholesome and uplifting choices for after-hours entertainment on our monthly weekend drills, usually at Camp Pickett.

Since these and many others were available from a movie rental company in Richmond – on 16mm sound film compatible with our military projectors – several of us chipped in to pay the modest rental fee. For awhile, Hollywood Horse Cavalry Movie Night at the C Troop pub was a big hit and a real morale booster to reinforce our unit's historical traditions.

But Boys Will Be Boys

That's not to say the novelty and morale value of those previously-mentioned "training films" lost favor. 8mm films had long been available and new short subjects in living color (but, sadly, no sound) were readily offered from the personal collections of many well-educated C Troopers.

Perhaps most notable in this vein is the night at Camp Pickett when no movie screen was available. Being daring, resourceful and well able to improvise, we just hung a bed sheet at the 2nd floor back wall of the barracks. Worked well enough but with a small, unforeseen problem.

A movie brightly projected on a thin white bed sheet can be seen not only from the front, but also from the back. No problem until some prissy MPs who were cruising the barracks area were attracted to bright, flickering lights and images "dancing" in the back doorway of the second floor. No doubt bored and wanting to bust up a friendly film festival, they climbed the stairs to shut down the "training."

None of the audience and exhibitors were arrested, but it's likely that our Troop's commander had a bit of explaining to do.

For the record, gentlemen from the more southernly geographical locations of the United States would never waste a single amp on silent B/W movies/training films.

Cornfield Detour
Carroll Childers, MG, ret

It was probably about 1972 and another MUTA-5 weekend drill assembly would start in a few hours. The Advanced Party had been assembled and briefed at the Dove Street Armory and were enroute to Camp Pickett to do what the Advanced Party does. Up until this trip I always practiced the concept that the Commander should be at the head of the Convoy or at least well forward in the march column. So there I was, actually driving a 2.5 ton Cargo Truck because the make up of the Advance Party was such that not enough drivers were licensed for the Truck so I was driving the truck and someone else was driving my M-151 Jeep. Maybe it was an M-38, I do not recall specifically.

We were on State route 153 which connects US Rt 360 west out of Richmond with US 460 West. 153 is much improved today but back then it was a typical narrow, very crooked road across farm and pasture country and lots of woodlands. I noticed in my rear view mirror a vehicle way back flashing its headlights which was a technique in convoying to signal a request for a convoy halt, used especially when there may be no radio contact between all vehicles. There certainly was no radio in the Cargo Truck.

I happened to be in a rather long straight stretch with a field on the right and easy access on and off the highway so I halted. The signaling vehicle pulled in behind me and it turned out to be one of our Advance Party in their POV, and there was no military convoy in sight behind him. The driver informed me that the convoy had halted a few miles back and that he had caught up to me. There had been a break in contact along the curved road and there may be a missing vehicle.

"Missing vehicle? What do you mean missing vehicle," I asked.
"That is all the info I have. I just left the convoy back there so I could catch you Captain" he replied.
"Well, I am not going to go back there in this two and a half ton. I will leave it here and ride back with you and figure out what is going on."
"Yes Sir, let me clean the seat off for you."

I turned the steering wheel hard right and locked it in place with a chain and boarded the POV for the ride back. No worry about the content of the truck as it was totally empty. Its mission for the day was to haul OEM after drawing it. Besides, nobody in their right mind would steal a US Army 2.5 ton empty cargo truck.

We arrived back at the convoy halt location and of course everyone was dismounted in a small cluster, hopefully discussing what might be wrong. As I got out to join them, Whitey Johnson came to salute me and offer the story. "Sir, I had been following this jeep then I realized that the next vehicle in front of me was not that jeep and there was a long gap between us so I knew something was not right. So I floored it and got the convoy stopped to count heads and that is all I know but I never saw the jeep turn off or anything."

"How far back did this happen Sgt Johnson?"
"Maybe a couple of miles. I was lucky with traffic and was able to get most of the convoy stopped."
"OK everybody, you all just wait here. Sgt Johnson, you and I will take this one jeep and that other jeep next to it, and we will backtrack and see what we can find."

So off we went, North on 153 to find the missing vehicle. By now they had figured out who the missing driver was and at this point in time, the name escapes me. He was a fairly new member but had been good about volunteering for this AP duty. About 2 miles North we approached a large corn field on our right and Whitey said "we are getting close because I remember this large field of corn now." He had hardly said this when the driver said "is that someone standing next to the road up there?" Whitey was elated "That's the driver!"

I instructed our driver to slow down and leave the pavement and get onto the grass strip between the corn and the ditch, which he did very deftly and we sort of ambled our way up to the soldier standing there looking like he did not know what to do.

We drove right up to the soldier, I dismounted as the jeep came to a halt; the driver with no vehicle in sight rendered me a salute, probably more out of fear than protocol, and I put him at ease.

"Where is your vehicle soldier?" He didn't take the At Ease; he was still at attention. He pointed to his left, "It is down that path through the corn, Sir, laying on its side."

"You mean you had a wreck?"

"No Sir, I think I went to sleep coming around that curve there", and he gestured to his right hand, "and next thing I knew there was corn everywhere, just beating the hell out of me; and the jeep went down through there and curved right and fell over and……."

I interrupted him to ask, "Are you hurt anywhere?"

"No Sir, I don't think I am hurt; maybe a scratch or two but I don't know if I was knocked out or not but I got up kind of dizzy and was lost in that corn at first and by the time I figured out to follow the jeep trail back to the road, the whole convoy had gone past and left me here."

"Sgt Johnson, check him out and tell me what you think. We will take him to HQ when we get to Pickett. The rest of us, let's go check on the vehicle."

There was a little bit of a ditch that he had plowed into and probably went slightly airborne then impacted in the corn field on all four and went on in 50-75 feet making an arc to the right, burying a left tire in soft soil and rolling over slowly and plopping on the left side totally hidden from the road. Cutting down all that corn slowed the vehicle down considerably, and, the convoy speed was probably only 45 or so when he left the road. No smell of gas or anything.

Well, let's all get a grip and set it back right side up and see if it will start. The venerable vehicle started right away and we continued the march to Pickett to complete the mission. The driver was uninjured and there was no way to know if he had been unconscious at all or just confused and scared.

That taught me to lead the Advanced Party from the rear.

Distinguished Honor Graduate of a VANG OCS Class Fails to Become TAG.

Carroll D. Childers, MG ret

The title becomes a question for prosperity. Why?

I was commissioned as an Armored Cavalry Officer in Class No. Six of the Virginia National Guard Officer Candidate School (OCS) in August 1964. After serving 32 years in the Virginia National Guard and commanding at every other level of command, he was selected to command the 29th Infantry Division. Three years later I retired due to having reached the administrative criteria known as Mandatory Retirement Date (MRD); an imaginary criteria related to the age of 62. Imaginary criteria is offered due to the fact that when 911 occurred, the MRD was almost immediately moved out to the right at least 3 years and in some cases, longer than 3 years to the right.

As a Major General (0-8), the maximum pay grade had been achieved but of course there was one more position in the state which carried what is arguably a more prestigious appointment. It is, however, a position of political appointment and the selectee serves at the pleasure of the Governor. The current Adjutant General (TAG) at the time I met my MRD, took office in January 1998 so that job would not be up for re-fill until 2002. Under the circumstances, I got on with the rest of my life and bid the National Guard a fond farewell.

On a Sunday morning in about April of 2010, I received a phone call from an old friend that I had not talked with in several years. Tom Berozowski was a wealthy and philanthropic individual with an intense interest in politics and whose support over the years to various key campaigns had gained him a special access and the associated influence to find himself in charge of some interesting political activities.

Tom's first words to me, after the gratuities of re-acquaintance, were, "I have been tasked by our Governor to provide him with a list of possible candidates for specific future appointments and one of them is for the position of TAG; would you be interested in that opportunity?"

My reply without a second thought was, "Absolutely, I would love that opportunity and cannot think of anything I would rather do. What do I have to do?"

"Based on what happened the last time you made this bid, and the circumstances surrounding that outcome, I would say that based on what I know now in this case, all you need to do is to apply," replied Tom.

"I do have one statement/question before I tell you about my current circumstance and that is, what about the current TAG? My impression is that he is doing a fine job."

"Well, I am not at liberty to discuss that but I can tell you that the current TAG will not be re-appointed," Tom replied.

"OK, then, whatever that is I suppose I will hear about it at some point so let me tell you about my current situation. My wife was diagnosed with cancer back in 2004, went through the surgery and treatment process and was declared "cancer free" as time went on. Then it returned as the cause for a cracked vertebrae. This too was treated and a monitoring effort instituted. About two months ago, complications with a hip led to detection of cancer as the root cause. She is undergoing treatment for this and the prognosis is good but as you can see, there is a pattern here which causes she and I to consider every decision with some care. The bottom line is that she and I will have to discuss this opportunity and get back with you."

"General I understand completely. Your first responsibility is to Dayle and you have certainly earned the choice of where to focus your availability and we know that one cannot be two places simultaneously. I will be praying for Dayle and if you can get back to me with the best decision for you and Dayle, we will carry on from there. Good talking to you again and look forward to talking later next week."

And talk we did. Dayle was up at the crack of noon and she recalled Tom from years ago. We had met Tom when an Army War College classmate of ours, COL Smith, was being presented the Silver Star as a result of the original recommendation, which had been lost in the shuffle, and he had discovered a process to revive the lost submission and the Board that approves such events finally did so. As I recall Tom and Smitty were on a county board of supervisors so Tom was invited as a guest and we developed a friendship.

The decision model chosen by Dayle and I was simple. It would simply be unfair to the NG for me to accept such a position of responsibility, at this point in time, knowing full well that if the cancer issue takes a significant turn, it could necessitate my stepping down. It was as simple as that. There was no way to know what the future held, nor what the timeline would be. Tuesday I called Tom to share the decision made by Dayle and I.

So I told Tom, adding, "But there is good news which I believe will offset any bad news you may perceive."
"I am open for good news General," Tom replied, "Got my pencil ready to copy."
"Alright Tom, write down Major General Daniel "Chip" Long phone number XXX-YYYY."
"Never heard of him," replied Tom.

"You never heard of him because he has been out of state, out of contact, serving America in other commands now for

5-6 years. He spent a year in Iraq as the senior contracting officer in charge of the 19 Billion Dollar reconstruction program following Desert Storm. Then he was NGB's point man Flag Officer to Hurricane Katrina. Then he was NGB point man Flag officer at the SW Border controlling illegal border crossing. Then he was made CG of a Joint Command at Ft Monroe and that 2 year assignment is coming to a close this summer. Chip Long is the most qualified, most capable, most adaptable, most creative, most likely to succeed General Officer in Virginia; or any other state for that matter. Plus, he is younger than I am by 8-10 years and he has a lovely wife who supports his every assignment."

"When can I talk to General Long."
"I will call him and brief him when we hang up and will give him your number and he will call you later today or tomorrow at latest."

So Tom and I end the call and he is clearly excited to "discover" MG Long and I am very pleased to have been able to play a role. I then call MG Long and tell him the whole story. He was quiet for a while after I relayed the story, implying of course that the opportunity was now his; the ball is in his court.

After a brief silence, he said, "Well, you know as well as I do that General Castles would expect one of us to take this job and do our best at it. I understand why you cannot commit to it. I had planned to do a lot of fishing after this Joint assignment, in fact, I have had a Bass Boat parked in my yard for 2 years and have never put it in the water. But if you cannot do the TAG job then I will do it. I will call Tom tomorrow after I have talked with Dianne."

And the rest is history. In the last of his four years as TAG, the Chief NGB was obliged to present 4 out of the 5

Annual NGB Awards to Virginia. I do not believe that level of performance has ever been recorded before.

My confidence in MG Long was fully justified. But I knew him well. He served as my XO at Brigade, followed me in Brigade Command, and oversaw the decommissioning of the Brigade due to Army reorganization. He served as my ADC (M) as well as the M for MG Blum and followed Blum in Command of the 29th ID. Then he did all those things which I articulated above to Tom. There was never a doubt that MG Long would excel. He was a RANGER and an Expert Infantryman. He was mentored by MG Castles, TAG-VA for 12 years.

Although I have no regrets to express, there is a point of history that may be worth accounting for. It has to do with an opportunity not taken, although, not taken by choice. Had I not been influenced by the circumstance described above, the VANGOCS would have had a Distinguished Honor Graduate who ascended the ranks to be not only a Division Commanding General but also to be The Adjutant General. The odds are very high that there will never again be so many independent variables aligning to all pierce the same target in complete unison as was the missed event described herein.

The second point of history that may be worth accounting for is the closely aligned record which MG Long did achieve. To date there have been 23 Commanding Generals of the 29th ID. Only one of these also became TAG; MG Daniel E. "Chip" Long.

Perhaps a challenge along these lines should be issued to the OCS Classes in the future and the recent past.

Tents Galore - Ft. Drum
LT Bryce Bugg

In 1972, in preparation for AT, Cpt. Childers gave all the O's and NCO's a list of gear and equipment that we were going to need -somehow get – at Camp Drum. I took notes, folded the list and stuffed it in my wallet. The next weekend, I visited my neighbor in Mathews County, MG Ciccolella, who just happened to be the assistant MFIC, US First Army. Told him we were off to Drum. He asked if we were short anything and I replied......

"Aw shucks, General Chick, a few things, but we'll get by."
"What do you mean by a few things?"
"Well sir, I just came by to say hello, but I do have a list."
"Gimme the list. Follow me."

He got on phone to the base commander at Drum and dictated the list.....and added and I quote: "Colonel, Lt. Bugg is a personal friend. I want him and the officers in his unit treated as VIP's and make damn sure they get that equipment."

Holy crap, what have I done? It's either gonna be real good or real, real bad. I called Cpt. Childers that night and told him what had happened. His reply was something along the lines of "making a grown man cry."

Upon our arrival at IGMR, I was greeted by the lovely and charming Sgt. L.P. Henning, who greeted me with... "Lt. Bugg you've really stirred up a hornet's nest!" It only got better.

Camp Drum and VARNG Maj. "Oggie" Ogburn, who greeted me with..."You ain't getting that %##&*."
"Ok," says I, "But General Ciccolella told me that if I didn't, he wanted to know who and why."

Oogie changed his tune and drove me to this enormous warehouse where every imaginable piece of equipment had a tag on it with my name. Bingo! Not only did we have more tents than Ringling Brothers, we had a permanent party detail to put them up. We lived well for two weeks and then a crew appeared to de-reckt, clean, fold, and return to storage with no paperwork. We had never even seen a GP Large tent, much less live in one and even park vehicles in them.

Reminisce
COL Ed Heavener

Thanks for sending the photos! Bill Atkeison and I were good friends back in those days, and I dug up his last known address after John contacted me. My wife Robin is Facebook friends with Bill so we reached out to him that way also...glad it produced some results.

The wedding in the photos occurred on 16 July 1977 in a little country church in Bath County near Hot Springs, and was attended by no less than four current and former Cav troopers, not counting myself. Ray "Rocket Man" Barlow was my best man, and Neil "Greasy Grass" Hening, Tony "Deerslayer" Burke attended along with Bill "Tin Indian" Atkeison. Neil, Tony, and Bill came in uniform, and Neil and Tony even held up cavalry sabers over Robin and I when we came out of the church!

A word about the nicknames. Bill Atkeison got the "Tin Indian" moniker because he always drove a big old "Indianhead" Pontiac car. Bill was really into photography back then and developed his own black and white prints. Bill was interested in technology too, and had the first Apple Macintosh computer I ever saw, which was quite primitive by today's standards with a black and white display, but it was amazing technology at that time. Bill worked for Maurice Harver in the Maintenance Section.

Neil Hening was a fan of the lore surrounding Custer's Last Stand, and adopted the name "Sergeant Greasy Grass" because "Greasy Grass" was what the Indians called the Little Big Horn River. Not sure about Tony Burke's "Deerslayer" nickname, but I assume it came from the James Fenimore Cooper novel of the same name. Both Neil and Tony were 2nd Platoon Scouts I think. One recollection I have involving these two troopers was attending Tony Burke's swanky wedding at the Country Club of Virginia, and the three of us drinking mixed drinks from that fancy bar out of canteen cups. We all thought it was pretty classy, but I think the bride (and her parents) were less impressed.

Ray Barlow and I were the unit ASTs from late 1974 to 1977, and since the Cav troop was a separate unit reporting directly to the 116th Brigade headquarters in Staunton, our workload was a lot heavier than other company size units which were organic to a battalion. Things like training schedules, requesting training areas, billeting, ranges, ordering ammo, rations and MTOE equipment were all unit responsibility since we didn't have a supporting battalion or squadron level S3 or S4 section to do the work for us. Ray and I worked a lot of late nights in the orderly room and were occasionally known to crack the seal on a bourbon bottle (after normal duty hours of course), since we were usually the only ones left in the building after hours.

Early Saturday morning of one our rare home station MUTA-4's, Neil Hening, Tony Burke and I were hanging out on the front stoop of Dove Street Armory before first formation when Ray Barlow pulled in after a particularly rough Friday night, driving a Corvette convertible he had recently bought from Johnny Mull. We all watched as Ray kind of crawled out of the low slung Corvette looking literally like death warmed over. The Elton John song was popular then, and Neil Hening yelled out "Rocket Man!!" giving Ray the nickname he would carry for the rest of his life. Sadly, the legendary Rocket Man passed away in August 2012 at the age of 63. My family and I were able to visit him in the months before his death, and my son and I attended his funeral service in Hillsville VA.

I have vivid memories of the Fort Pickett tank incident recounted by MG Childers. Most of the time us headquarters types didn't get to do much of the fun stuff, but that night Barlow and I somehow talked SFC Talbott into taking us along for the ride in his tank. Our driver was a prior service trooper named Robert Cutchens who had tank driving experience in Germany. The driver was buttoned up following the cat-eyes of the tank ahead, Barlow and I were in the gunner and loader seats, and Talbott was standing in the TC hatch. We had been underway for a couple of hours, and since we couldn't see anything from inside, I was dozing on and off, but I remember hearing Talbott telling Cutchens several times that he was running off the trail to the right. (Later we heard that his optics were misaligned so that Cutchens thought he was going straight when he wasn't)

All of a sudden I heard Talbott yelling, "STOP! STOP!" along with the sound of wood splintering! The tank stopped and we carefully crawled out of the turret and down off the tank's left side. Exiting the hatch I remember looking down over the right side of the tank and seeing nothing but the Nottoway River below. We had taken out at least 15-20 feet of guardrail, and about the front third of the track was out over the water, while the rest of it had climbed up on the side rail of the bridge. We all assumed the M88 would have to pull the tank back off the bridge, but SFC Harver rolled up and insisted on driving it off, which he did with unbelievable skill. As MG Childers said, this was a very close call, but SFC Talbott's alertness along with the quick reactions of Cutchens to stop the tank was all that saved us that night.

(I pronounce this one to be the most serious story in the collection, MG Childers)

The Human Flame Thrower Affair
Carroll Childers, MG, ret

Do not try this at home; or anywhere. It is not safe, especially if you have inhibited your reflexes and your sense of executing specific steps in the proper order by ingesting substances known to diminish clear thinking.

C Troop 223rd Cavalry was training at the State Military Reservation (SMR) in 1968, the first annual training period they had conducted following the retirement of the 29th Division. James A. Bailey was the C Troop commander, with Joe Lucas his XO; Dennis Pounds, Buddy Deverell, and I were the three Line Platoon Leaders.

Capt Bailey was a highly social animal and took every opportunity to attend functions wherein there was food, drink, and opportunity to interact with others. He particularly enjoyed mingling with senior officers but perhaps no more so than he reveled in pulling some trick on peers and juniors to effect some level of embarrassment or to put them in a position where they had no reasonable response.

I think it was on Saturday, middle weekend of the Annual Training, when Bailey summoned his 4 junior officers together to announce that we were going to attend a social function being held by a Battalion sized command, also at the SMR. Bailey announced the uniform (khaki with Cavalry neck scarf and web belt with holster and sidearm (no ammo), boots bloused) and time to reassemble in his office for movement to the affair. We were in the field uniform and did not have much time to prepare for such an affair but orders were orders.

We reassembled on order and Bailey was sitting with his boots on the table, smoking his pipe, with a giant grin on his face. "OK here is what we are going to do," he announced. "Childers, I have seen you do that flame thrower trick so here is your fuel and a lighter. We are going to go over to the (redacted unit name) where they should be well into their social. I am going to kick the door open and LT Childers will be the first one in to flame them out then we will proceed to eat them out of house and home and match them drink for drink. Any questions?"

Nobody had a question so off we went. We marched a few buildings behind Bailey and as I recall entered an old mess hall where the social affair was underway. Sure enough, Bailey thrust the screen door open and it banged against the handrails as I recall. I had taken a deep breath, held my breath, squirted an ounce or so of Ronsen Lighter Fluid in my mouth, stepped over the threshold, forcefully expelled the lighter fluid as a fine mist, while lighting it with the lighter. I would say it was a good 5 footer; maybe the best one I ever did. Bailey stepped around me and made some proclamation which I do not recall as I needed to step back out the door and rinse my mouth out with a coke I had brought for the occasion. Bailey was inside attempting to turn the gross interruption into some kind of occasion of joy or bewilderment. Tension in the air was about as comfortable as would be tear gas when you failed to bring your protective mask. We began to mingle and consume their food and toxic drink until they eventually ran out and we left.

One attendee ("host" unit) held this incident as a grudge against me for over 20 years and delayed my becoming a 0-8 for about 3 years. Some folks can't take a joke.

Getting Fired by the Best
Carroll Childers, MG, ret

The precise year of this story is not guaranteed but it must have been about 1973. I was the C Troop commander of C Troop, 1-223rd Armored Cavalry Squadron, with Headquarters in Philadelphia PA. C Troop was located in Richmond VA and we were all Virginia Guardsmen but because of a reorganization in early 1968 with the retirement of the 29th ID and the loss of an entire Cavalry Squadron (the 1-183rd), a deal was cut between the Adjutant Generals of PA and VA to place one Troop in VA. When the 29th was retired, I lost my assignment as a Platoon Leader in C Troop, 1-183 in Fredericksburg but someone was looking out for me and when I returned from Viet Nam in the summer of 1968, I found that an assignment for me had been saved in the only remaining Cavalry Troop in Virginia.

There is no way to describe my joy to learn that I was given a slot and the great pleasure and thrill of continuing as a Cavalry Platoon leader. This assignment continued for a couple of years and I was promoted to the Executive Officer position replacing Joe Lucas. A year or so of this and it was time for the C Troop commander Jim Bailey to move on which elevated me to C Troop commander.

My entire career in the Guard at this point had been in Armor, but for one year in an Air Defense unit while waiting to enter OCS. Making C Troop commander was by my own estimate next to being crowned. I set about the challenge at full throttle and with all the professionalism I could bring to bear based on observations of leaders that I admired, the lessons of OCS, and those troop matters that I had absorbed as an enlisted man for 9 years.

At this point in my career I had a relatively small baseline of other leaders on which to attempt to craft the model of a leader that I would attempt to become. First was my Tank Company Commander, Captain Bob Downey, a WWII and Korean War veteran Infantryman. Here I had been a First Cook for less than three years before being reassigned as a Tank Driver for several years. That transition in career paths is the subject of another story. Second was the most instructive of all, a WWII soldier named Alvin York Bandy who had been my C Troop commander in a Cavalry Troop in Fredericksburg and who had allowed me to enter OCS. Captain Bandy was my C Troop commander 63-68, was my "life-long" mentor, and greatly sculpted the soldier that I would become. Finally there was Captain Jim Bailey who was my C Troop commander from 68-71. Of course there were several other Cavalry Officers in the 1-183rd Squadron which I observed and adopted some characteristics from but they were more distant and offered less direct influence on me. Of these, Captain Mike Marshall is the most memorable.

C Troop, 1-223rd, began life as a very select organization, literally, because it was formed by selective manning from the entire Squadron; three line Troops and a HQ Troop reduced to a single Line Troop. It began as a slightly over-strength unit and of course did not take long, with retirements, to draw down to a full strength unit. Then began the challenge of maintaining strength in the light of the National and world situation. Viet Nam was underway and protests were getting far too popular. Recruiting and retention challenges was the greatest of all others combined.

C Troop was training hard, having lots of fun, and catching the attention of higher headquarters, which was the 116th Infantry Regiment whose HQ was in Staunton. The Regiment, responsible for oversight of our training and administration, were regularly sending staff and other observers to our trainings to evaluate and report accordingly. They began to discover a Company-sized unit with good leaders and NCOs and a body of enlisted soldiers who were extremely knowledgeable and capable and who also exhibited esprit and a unique can-do attitude. Pleased with what they saw, the 116th Staff chose to invite a team from C Troop to accompany them to Ft. Bragg in support of a training exercise known as a Command Post Exercise (CPX).

The CPX was a rudimentary version of the modern day "Warfighter Exercise", but was far more manual than the Warfighter which is heavily supported by computer programs. The Brigade also recognized that C Troop was not adverse to hard work and could be relied on to take on any task with great relish whether or not it was erecting tents or playing tactical roles as part of the graded exercise. Of course, C Troop thought that they knew far more about tactics and techniques than any grunt could possibly know, after all, at the platoon level in Cavalry, soldiers must know how to integrate and employ reconnaissance, Armor, Infantry and indirect fires in a tactically sound manner.

So there we were; we had convoyed down to Ft. Bragg from Richmond, linked up with the Brigade in the field, and did the greater portion of setting up exercise tents and field accommodations for spending some 10 days there. This included not only CPS tents and field latrines, but generators, lighting, sleeping accommodations, parking, communications, security, overhead camouflage nets, and food service support.

So now I must divert to another portion of the story and then bring it back to this point to wrap it up.

Mentioned earlier was the challenge of maintaining strength. Our Troop strength was down approaching 80% and in meetings with the troops, the story was that grooming standards in C Troop were negatively impacting our numbers and that was a primary factor. Civilians had a propensity for long hair (compared to military) and facial hair and the implication was that if some accommodations could be allowed in this regard that the strength issue would improve. With great reluctance I agreed that a suggested approach known as a "short hair wig" would be allowed. The short hair wig would allow the concealment of actual hair, too long for military standards, beneath the short hair wig. The deal was that C Troop would tolerate short hair wigs for one year and if strength improved then the short hair wig could continue and concurrently, if strength did not improve into the 90's minimum, then the short hair wig tolerance would end.

A year went by and the strength did not break 90% although it was close. I drafted a letter and distributed it to all members announcing the end of the short hair wig trial. Naturally this did not set well with all members and two soldiers in particular went to the Adjutant General and registered a complaint. The Adjutant General at the time was a Major General who had found some disfavor with me way back when I graduated OCS and became a Second Lieutenant. I always felt that he was disappointed in a decision that I had made relative to my first post-commissioning assignment and now, with the soldier complaint, he felt he was in a position to retaliate.

This was, as I initially guessed, about 1973 and communications was not what it is today; no email, no texting, no instant communications. What was it? Telefaxing from one headquarters to another and then courier in a jeep to the field to deliver the message. And so the courier arrived at the field Command Post (CP) with about 3 days left in the exercise. The letter sent by the Adjutant General to the Regimental Commander basically instructed the Colonel to validate that I had sent a specific letter to the Troops "showing great indiscretion which could not be tolerated in a Commander." If I had in fact originated this letter, I was to be dismissed from the exercise, returned to home station, and instructed that I would be separated from the National Guard.

I was in the CP enjoying the exercise and the team from C Troop was doing a great job and making remarkable contributions to the CPX to the extent that they were the discussion around the water cooler so to speak. I was handed a note to report to the Commander's tent. I immediately walked the 100 feet or so to his tent, saluted, reported, and was told to stand at ease. He was nervously puffing on a cigarette and was clearly upset about something. He finished his smoke, snuffed out the butt. He was sitting in a field chair at his field table. He picked up a sheet of paper turned upside down and flipped it from his hand to my side of the table. "Did you write this letter Captain Childers?"

"I don't know yet Colonel, may I look at it?" He gestured with his palm up indicating for me to do so.

I looked at a poor copy of something that looked like the original may have been wadded up then later refurbished to some extent, then copied. "Yes Sir, I did write this," I replied to him.

"In that case," he responded, "I have been instructed by the Adjutant General to inform you that you are to proceed back to Richmond immediately, pack out your desk and office, turn in your issued equipment, and consider yourself separated from the Virginia National Guard. Instructions will follow on details of a change of command inventory. Questions?"

I saluted, did an about face, and marched out and back to the CP to find my number two and relate the situation. Then I drove to main post, found a phone booth, and called my old TC Alvin Bandy and told him the story.

Alvin was a guy who never showed concern about anything. He always had another option to play. "Don't worry about it. Continue the march and follow their instructions until someone tells you differently. I will call Jack Castles and see if he can calm it down."

Castles was a COL at the time and held the position of Chief Of Staff to the Adjutant General. Castles later became the Adjutant General himself and served three different Governors in that capacity. He apparently contacted the offended Adjutant General and convinced him that "firing Childers is not the right thing to do."

It was several hours later that I arrived at the Dove Street Armory in Richmond. I contacted Bandy again and was relieved to find that the firing was out. No need to clear out my office.

Grounding the Aviation Fleet
Carroll Childers, MG, ret

Much like the blowing of the field latrine, C Troop managed to ground the entire aviation fleet by a single act of tactical genius.

C Troop was not the only unit that enjoyed playing pranks on others. All other units were, perhaps, timid in comparison to the CAV and the Aviation and so it became more of a challenge for us to attempt to get the best of each other than it was for either of us to waste our time on the timid brothers in arms. And so we became rivals in the field and in the Cantonment area. Such special activities as drinking beer, mooning, display of special paraphernalia, risky moves, display of strength, inventing special names for each other, booby-trapping, trashing rooms (when we were in barracks), etc etc. Nothing much was off limits. The aviators were almost entirely veterans of Vietnam. Some had been gunship pilots, some had flown or crewed slicks and medivacs, some had been cargo flyers. In contrast, very few of Cavalry members, and especially the officers, had combat or combat zone experience. We had a few enlisted men who had been in Vietnam and I was the only officer that had been in Vietnam, and I as a civilian science advisor and not as an official combatant. But somehow the Aviators sort of accepted my 8 months in Vietnam as creditable service.

Aviators had several advantages over the Ground CAV. First, they were fast and could be there one minute and gone the next, with only a few seconds exposure and even then, we had no real access to them to do them dirty. And of course the idea was not to do anything that might make their chopper fall out of the sky so there was very little we could do except to try to get even after the fact for what they might do to us.

Having been in Vietnam as a science advisor who introduced several wonderful items into service there, it seemed like sometimes found myself with a few items left over in the development process. On one occasion, I had some aerial flares that were fired from a 12 gauge shotgun. I took some of these to annual training for a completely legitimate purpose of signaling. One thing the aviators seemed to enjoy was to come and hover low over my CP in hopes that the downwash would disturb, if not completely blow away, my tent or the extension out the back of the command track. So here they came one day to hover over my CP, which was set up under some trees. They brought the chopper right down with the skids in the trees and there was quite a windstorm on the ground. I took a flare round and shot it into the arc of the tail rotor. This was not done haphazardly, rather, being an engineer and an expert in ballistics and materials experimentation, I knew that the launched element of the flare round was little more than a soft glowing ember with zero capability to cause impact damage to material and insufficient thermal capacity to create ignition. These were blown away harmlessly by the rotor wash. Naturally it hit a blade and exploded into little sparks in a great shower. This scared the living Poo-Poo out of the pilot and crew as they did not know what had happened and they had never seen anything like that before. They "pulled pitch" as they say and got the hell out of there post haste.

They felt that they had been taken advantage of in some unknown way. So they made up a false aviation incident report and sent a three-man team over acting like some sort of official incident inquiry team. It did worry me at first, but as the inquiry went along, I began to detect typical aviation BS and I told them that I could not proceed with cooperation until I had gathered my witnesses to the unsafe and un-requested hovering above our CP. What if they had a power failure with their skids in the trees directly above my encampment? There would be no way to recover and the chopper would fall on a manned CP. I would never tell them about the 12 gauge flare round. Eventually, it all ended in a tie. They never hovered over my CP again, but this is the only instance where we were able to take any action against the Aviators in the air.

One day the Aviators were particularly obnoxious to the CAV. We were moving in column formation along a tank trail when they swooped in low and bombarded us with all manner of plastic bags full of things from the air. Only they knew the full contents of the bombs as many of them missed vehicles. I know there were hits with flour and with eggs and with what could have been the contents from their mess hall grease trap and garbage cans. This was not the first time they had hit us with flour sacks but the other styles were different and caused us to reconsider normal retribution, which was to paint things on their aircraft. This time, we would give them something to remember.

We were playing tactical games that included the Aviation. Their LZs were tactical and were relocated at unpredictable intervals but we found out through surreptitious means that they planned to move back to the airfield for one night as a "tactical diversion." Actually it was their way of getting back in for a shower.

Our objective would be to steal the log books out of the aircraft, thus grounding them. By regulation, they can't fly without the log book. They can do a lot of silly things, but violating aviation regulations is where they generally drew the line. The Cavalry Troop was made up of 3 line Platoons, each with a scout section mounted in jeeps with 30 caliber pedestal mounted machine guns. In addition, each platoon had a rifle squad normally mounted in M-113 personnel carriers. The plan was to cause a diversion by having the three Scout Sections drive through the hangar and billeting areas around the airfield blasting with machine guns (firing blanks) and tossing simulators and smoke grenades while the dismounted infantry made a raid on the line of helicopters, taking the log books from the cockpits where they were normally kept after crew maintenance. The Scouts created quite a commotion, and so furious were the aviators that they called the city of Blackstone and asked that the police respond to the "dangerous activity" by these people in gun jeeps. By the time the sirens and flashing lights showed up, the scouts and infantry had long gone...along with all of the aviation company logbooks.

About 0700 the next morning I had a visit from the Squadron Commander, his XO, his JAG officer, and the Aviation Company Commander. I invited them to join us in breakfast.

In the sternest manner I had yet seen from the Squadron Commander, he replied "This is not a social call Captain Childers. Did you take the Aviation Company log books last night?"

I replied, "Sir, I can assure you that I did not take anybody's logbooks. What would I possibly want anybody's log books for?"

The Commander responded, "You know that they cannot fly without log books?"

"I had not really thought about it that way sir. Let me look into it. It could be that some of the men decided to borrow the log books to examine them for some evidence of the Aviation having illegally thrown stuff out of the sky at our column, including simulators. What if one of those simulators had gone down a tank hatch? Fortunately none did, but a lot of pretty nasty stuff did hit our vehicles."

The aviation commander is beginning to squirm a bit now because he realizes that this safety issue is serious and he wants to get out of this debate. He offered, "Sir, if I can just get my log books back so I can continue training I am willing to forget the whole thing, including the simulators in the tents last night."

"I do not have the log books sir, but I will make an immediate inquiry and I will do my best to locate them or help the aviation commander recover them."

"You do that Captain Childers; NLT 0900 or you are in deep Kimchee."

They left. I summoned my First Sergeant and instructed him to collect the log books, put them all in a box or bag, take them to some location where they can be found; then call his counterpart in the Aviation Company and tell him where the log books are. No person to person hand off. Make it into a training event. Have the scouts involved to provide overwatch and gather information on who recovers the log books, time of recovery, bumper number of vehicles involved, etc; and to infiltrate the airfield and place their CP under surveillance via field glasses and report arrival of log books; and to trail the vehicle that recovers the logbooks. I wanted evidence that they got their books back.

We got even. End of story.

The original M-48A1 Main Battle Tank with M41 90mm Main Gun.

M113 Armored Personnel Carrier for the infantry dismounts.

Early M-38 Willys Scout Jeep with .30 Browning M1919 MG.

M88 Armored Recovery Vehicle.

Interior of the M106 Mortar Carrier with the M30 4.2in Heavy Mortar.

M577 Armored Command Vehicle.

M48A5. Last of the M48s in service (1975) with the L107 105mm main gun.

.50 caliber Browning M2 heavy machine gun.

7.62mm NATO M-60 machine gun in armored / aviation configuration.

11B Infantry and Mortarmen on an M113 with M16A1 rifles.

Cav on the move in their natural environment, the tank roads of Ft. Pickett.

Tank trails and creeks, Ft. Pickett, VA.

HQ in the field, Ft. Pickett. Later issue M151 Jeeps.

Main gun range, Ft. Pickett, VA.

Main gun range, Ft. Pickett, VA.

Negotiating a tank ditch, Ft. Pickett, VA.

Infantry weapons range, Ft. Pickett, VA.

Rappel training at the quarry, Ft. Pickett, VA.

Air MEDEVAC on the training ground, Ft. Pickett, VA.

Another day, another thrown track.

M88 to the rescue, again.

11Bs waiting on orders.

Lined up at the wash point. Ft. Pickett, VA.

The motor pool's job never ends. Engine swap out.

Guarding the Mess Hall
Advance Party, 1982 Fort Bragg
MSG Bill Talbot

The AT Advance Party arrived two days ahead of the main body of the C Troop convoy. The advance group was to report to the post and begin drawing buildings and all the equipment needed for two weeks of training. First Sergeant Bill Talbott led the group.

As background, this was C Troop's first AT training at Bragg so the advance party group was dealing with a new group of Bragg people in charge of the post buildings we'd be using as well as the tanks and other armored vehicles needed for the training period. Only someone who has served on the advance party for AT or a MUTA Five can appreciate all the gear that has to be drawn from the training post. Machine guns ("shootin' irons" as Lt. Bugg calls them), and commo gear, mortar tubes, all sorts of maintenance gear, mess equipment and on and on.

First Sgt. Talbott and a several NCO members of the advance detail drew our assigned three barracks, the orderly/supply building and an adjacent mess hall while others were signing out for tanks, mortar carriers, and M-113's for use by each platoon's scout and rifle section sections, plus an M-88 tank retriever, a lighter-duty wrecker truck, and a fuel truck.

When Talbott arrived at the assigned company area, the keys he had been given did not fit the locks on the orderly/supply room or any of the three platoon barracks buildings. The only key that worked was the mess hall key. Talbott needed to get the keys that fit the other buildings.

While he and his crew returned to the post officials to get the right keys, Talbott ordered two deuce and a halfs loaded with M-16's and other C-Troop gear hauled from Richmond to be off-loaded for safe keeping in the mess hall; the only building he could gain access to.

Before leaving the company area to fix the key mix up, Talbott assigned a young soldier to stay behind to keep watch on our gear. Talbott told the guard, "Son, whatever you do, don't let anyone in that mess hall until I get back. Do you understand?" "Yes, First Sergeant!"

An hour later, Talbott was still at the post billet office when two other advance party members found him and said, "Sarge, you better come quick, Private Doe won't let us in the mess hall and he's waving a butcher knife at anyone who comes close!"

Rushing back to the company area, they found the young guard standing at the mess hall door, butcher knife in hand. Thinking the young man was high on something, Talbott calmly told him it was okay and to put the knife away...all was well.

Turns out the soldier was high...on caffeine. He'd found a packet of iced tea concentrate, meant to make a five-gallon batch, and had used up the whole packet mixing a cup at a time to drink while standing guard.

Talbott recalls Trooper Doe being on "full alert" until the following Tuesday.

IG Inspection
John Terry

"You don't pull on Superman's cape, you don't piss into the wind, you don't pull the mask off the ole Lone Ranger, and you don't mess around with Cav!"

Sometime in the '76/'77 time frame the unit was undergoing an IG inspection. I believe it was Capt. Deverell's first. In the morning walk through the inspector noticed a blank space on one of the metal shelves in the supply room. Upon returning to count various parts he noticed that the space now had radiac meters. He immediately challenged the appearance of the equipment. Lt. Redfern (XO) explained that the meters had been at OMS for calibration and that another AST was picking up their inventory and brought them back for C Troop. The inspector didn't buy that and challenged the Lt's integrity. At which time Lt Redfern being a Southern Gentleman and Cavalry Officer took great offense at the implied assault on his honor. This irritation continued all morning.

A little after lunch both Maj Flower and I showed up to lend moral support. Maj Flower was an active duty officer assigned to Readiness Group Ft. Lee. While he had responsibility for several of the Brigade's units, he took particular interest in C Troop. Since he was a Calvary officer and had served with the 11th ACR in Nam.

The inspection had moved to the orderly room and the inspector was seated behind one of the AST's desks. You may remember it was customary to have several large 3 ring binders standing in a row on the front of the desk. These binders contained many of the working papers, manuals and rosters that the AST's worked with daily and allowed for quick access rather than having to get up every time they needed a manual or regulation.

Now this is important, because what I am about to relay could not have happened without those binders stacked on end in a neat row.

The inspector had just returned with a fresh newly brewed cup of coffee. Without even milk to cool the coffee, he set the piping hot cup down right in front of him and proceeded to sit and pull his chair up to the desk. With a quick moment Lt Redfern dropped, to this day I don't know what, something on the corner of the desk. This caused a vibration which in turn caused the first binder to tip to its right. Hitting the next binder and causing it to tip over and hit the next binder and so forth until the last binder fell flat and caused a stronger vibration than the original caused by the Lt.

The desk moved ever so slightly but enough to cause the coffee cup to rise up in the air and then proceed to tip over into the inspector's groin. He immediately jumped straight up as if propelled by a rocket up his butt. Standing, with tears in his eyes and steam rolling from his pants. Frankly I gained some respect for the man in that he did not curse or cry out in pain after the initial exposure to the scalding hot coffee.

Now several reactions occurred at this time. Lt Redfern was searching for something/anything to wipe up the coffee in the groin of the inspector. Fortunately he couldn't find anything immediately. Can you image a Lt touching the groin area of the inspector? I can't either. Probably best that he didn't touch his groin. Capt. Deverell almost went into cardiac arrest. His eyes appeared to triple in size and he seemed unable to speak. Maj Flower and I looked at each other and immediately started to laugh. With that we walked at a quick pace to the doors, turned right and ran up the hall. At a reasonable distance from the orderly room we stopped and laughed until I thought we would be sick.

After regaining our composure, we made our way back to the orderly room where everyone tried to act as if nothing had happened. The inspector still had red eyes, but continued on professionally. I must say a better man than I!

Somehow the unit passed inspection. For many years I figured that the unit passed because we deserved it. But after some thought, I have come to a different conclusion. Lt Redfern had clearly sent a message. And that message was received, loud and clear.

Don't mess with the Cav!

A Knife at a Poker Game
Alan Hughes

The game started around 8 P.M. with the usual players plus a big fellow from another guard unit who was making up his missed AT with C-Troop. We are playing in the mess hall, our preferred location.

Around 10 P.M. another fill-in soldier (we'll call him Joe) came in with a small knife in his hand and looking pretty inebriated. My first thought was that this small guy, Joe, wants to take on five guys and rob us. Cy Coleman, our commo section chief and a regular at the table, picked up a chair to defend himself. Just then, the big fellow sitting in the game grabs Joe and squeezes the air out of him. Joe drops his knife and runs out the door.

We lock the door and continue the game. Two hours later, Joe is back, crashing through a mess hall window, getting cut in the process and was bleeding profusely. His friends take him away to get medical treatment. Again, Joe and his buddies were all assigned to us to make up for missing their unit's AT. Even this second intrusion did not put a damper on the game. Someone deal! The game ends at midnight.

At 3 AM, Cy is awakened by a loud tapping sound. Joe was standing over Cy's bunk, tapping the end of a baseball bat on the floor. Loud words are exchanged alerting Joe's friends who come and convince him to leave. Why does Joe want to harm Cy? As Paul Harvey would say - stand by for the rest of the story. Here goes:

Captain Childers assigned Cy to be the captain of our softball team. Our team practiced and Cy picked out the lineup. Joe was not picked as a starter which did not sit well with him. Being slighted in the team selection was what drove Joe to get even with Cy with his anger apparently fueled by multiple brewskis.

Whoops!
* Redacted *

Location Office Park South Richmond

An Armored Cav enlisted Soldier and his High School Friend were cruising in a 1971 Ford on Belt Blvd drinking Schlitz Malt Liquor and looking for female companionship. While working on the second 6 pack of Schlitz – (currently under FDA review), The Armored Cav Soldier (ACS) confessed to his friend that he had just returned from his National Guard Annual Training and he had smuggled some Artillery Simulators home with him.

His friend, who by the way was a sweet innocent naïve individual until he started hanging out with his gangster buddy, had no idea what an Artillery Simulator was – much less what it did started to ask the ACS about said object.

After much discussion and many more cans of Schlitz the ACS agreed to a demonstration.

Parked by an office with a glass front window The ACS handed the simulator to his friend and said all you have to do is pull the string and throw. To the horror of the ACS -- his friend, who is not known to have a strong base of common sense, (hence a strong prospect for Officer Candidate School later in life) ACTUALLY PULLED THE FREAKIN STRING AND ACTVATED THE DAMN THING – The ACS told his friend to throw it out of the car. The simulator hit the ground – bounced a few times and landed right in front of the pane glass store front.

As the simulator started to whistle the ACS muttered "HOLY SHIT" – put the Maverick into gear and spun out of the parking lot. Before they had reached the exit the simulator exploded and the shock wave caused the buildings front glass to shatter – crashing to the ground.

Two scared shitless young men were traveling West, quickly, when about 5 minutes later there were a plethora of state and local police heading East passing them at a high rate of speed – Blue lights flashing and sirens pitching the otherwise quiet moonless sky.

Those involved would like to think that with all the years past all would be forgotten and forgiven. Yet sometimes you will hear some old retired state trooper tell the tale of the unsolved cold case of the first pre- 9/11 terrorist attack happening on the South Side of Richmond, Virginia.

END OF NARRATIVE – I will not confirm nor deny if the above events are true nor any knowledge of the individuals involved in alleged incident.

Incoming! Shooting outside the lines, Ft. Pickett VA.
Carroll Childers, MG, ret

It was likely 1973 because we were conducting Annual Training at Pickett as we did on odd years. Time goes by when you are having fun they say. I was the Troop Commander of C Troop, 1- 223rd Armored Cavalry, Virginia Army National Guard. We were at Annual Training at Ft Pickett VA. It was Saturday morning and a lot of nothing was going on. Back then, when the Army had less interest in the National Guard than later, our practice was to train hard for the first week, conduct some maintenance and administrative work till about 1200 or 1300 on Saturday, then give the troops a pass until Sunday night. Then come Monday morning, we would go back to a rigid training schedule through Thursday; then Friday we would do the maintenance required to fix everything we had broken over the previous two weeks in preparation for returning to home station on Saturday.

The National Guard was always provided with Active Component Advisors, both enlisted and Officer. Usually we had a full time senior NCO who was available 7 days a week during the year and when we went to Annual Training (AT), we would get a Senior Army Evaluator to observe our training and inspect everything we did. We would get a grade on AT performance. During this particular year, we were assigned a Major of Infantry named Tony Le'Heur. Tony was a combat infantryman with two tours in Viet Nam; one as a platoon leader and one as an advisor to the Vietnamese Army, specifically, a component known as Regional Force/Provincial Forces. They were referred to as RUFF-PUFFS.

I was in the orderly room doing some officer evaluations and other admin work and Tony came in to check on things. He knew I had spent some time in Vietnam and wanted to talk about that some. He told a story about his assignment with the RUFF-PUFFS. He was visiting a village one day and was squatting in the Vietnamese fashion eating fish and rice with the village elders when a patrol returned to the village. They were very excited chattering about the ambush they had successfully pulled off against some NVA. One of them was carrying a duffle bag. Tony could speak the language and understood what they were talking about. The village elder then asked to see:" Show me! Show me!"

The patrol opened the duffle bag and rolled out 6 severed heads and began to kick them around and play with them in various ways. Tony said, "I had enough fish and rice."

He had hardly finished the story, was up getting a cup of coffee in preparation for perhaps another tale when all of a sudden there was a swishing noise with a hint of a whistle outside somewhere. Whatever it was it was moving fast. Without any other thought or action, he dropped the cup of coffee and sprinted across the room for the telephone and dialed Range Control.

"Range Control, can I......"
"Cease Fire! Cease Fire ! Close all firing points now!" shouted Tony, obviously interrupting the range control NCO.
"Who is this?"

We heard an explosion some distance away.

"I said cease fire. You don't ask why when a cease fire is called for you just do it; now do it. I am on my way over there. Have your Colonel on hand!"

About this time the second swishing sound went over. He looked up at the ceiling. "Number two just went over my head in the cantonment area. Cease fire. Out."

The second explosion occurred. By now, we had all figured out that two rounds of artillery had passed over our building and would impact somewhere outside of the impact area. A dangerous situation, and on my side of the Post. Tony ran for his vehicle and was last seen heading towards the Range Control and a showdown. I ran for my jeep with driver in tow and headed towards the sounds of the explosion. We had no idea where the rounds landed.

As I got underway, I turned on my tactical radios. It was this year that I had decided to do something unusual with my vehicle. The normal configuration of jeep and radios is to have two radio racks, one on each side of the vehicle adjacent to the back seat. To operate the radios from the right front seat, I would have to turn and twist to access the radio controls. So this year, I had taken out the front passenger seat so that I could sit in the back seat between the two radio banks, then, with the seat missing, I could get in and out of the vehicle very easily and quickly. This year also, I had installed twice the standard numbers of radios so that I could constantly monitor numerous nets without having to change frequencies back and forth.

So immediately put out a command call asking for information on possible incoming. Before I got more than a few blocks, I got a call from one of my units at the ball field that the two rounds had impacted there. Nobody was hurt but when I got there, the game was suspended and everyone was gathered around the club truck drinking beer and telling tales about the two errant rounds. The first hit in the edge of the right outfield and the second fell shorter and impacted just between the outfield and the paved road that bordered a barracks building.

Major Le'Heur returned from the Range Control somewhat calmed down but still fuming and somewhat triumphant having had a piece of the Range Control for their initial response. An Investigation ensued as to how this firing out of sector could possibly occur. The outcome was a simple explanation. The Artillery firing unit had gone through all of the steps to gain permission to be firing. They were practicing shoot and scoot missions wherein they would pull into a firing position, respond to a call for fire, fire the mission, then convoy toward a subsequent firing position simulating typical fire support. Then when they approached another approved firing point, they would simulate another call for fire support, pull off into the pre-surveyed firing position, and provide fire support. This was all done by a fire support plan and overlay that had been approved a week or so earlier.

When the Range Control Officer signed approval of the plan, he would initial adjacent to each preplanned firing point approving the firing data written there. On this particular firing point, his initials were carelessly placed so that his pen obscured part of the firing data and when the gunner set the gun data, he dialed in an incorrect azimuth, thus the gun was pointed out of safe sector and towards my ballpark.

Thanks to an automatic reaction by Major Le'Heur, who had plenty of experience in hearing live artillery go overhead, and his no nonsense response to the lackadaisical range control NCO, I believe that he probably prevented some amount of damage, wounding, or death that day.

This was my second experience with similar errors. It was about 1966 and again it was at Ft Pickett. I was a Lieutenant, Platoon Leader, and we were in the tank parking lot at the end of a day of training doing maintenance on M41 tanks. Rounds began to impact IN THE PARKING LOT.

We immediately got behind tanks. The rounds were coming in about 100 yards from us and it was a fire for effect. So happened that an MP car was passing down the street and called range control and demanded a cease fire. We had a number of fragments ricochet off of the outer line of our tanks but no one was injured. Believe me, the Artillery Regiment was harangued for years for having made that little error. There was never an explanation given for their error.

Integrity
SPC-5 Mike Thomas

William K. Jones (Billy) Jones was a regular Army veteran, airborne qualified, who worked full-time as an armorer for the Virginia Army National Guard at Camp Pickett. (Some soldiers took the extra effort to call him "William K.

Billy came to be a member of C-Troop in 1971, serving as a Tank Commander and Platoon Sergeant in the third platoon. He was immediately recognized and welcomed as an experienced hand. He was looked up to and trusted by our officers rank-and-file alike. Billy believed that everyone had something to contribute and he treated everyone with respect. William K. Jones made C-Troop better just by his presence.

In 1973, our Annual Training was at Pickett. (We rotated to Drum in even years, and Pickett in odd years.) A and B Troops were also on hand for squadron level exercises. Somewhere on Pickett's miles of tank trails, someone lost his wallet. William K. found it. The ID in the wallet said the owner was a fellow who lived in the Philadelphia area, A-Troop's Headquarters. Billy passed the word up the leadership ladder that he was looking for "John Doe" of A-Troop.

Finding a wallet in the middle of a Pickett tank trail was a miracle in itself. That Billy Jones was finder was even more important. There was no doubt that the wallet, and the considerable amount of cash it contained, was going to be returned to the rightful owner.

A meeting was arranged, in the field, with the A-Trooper and his platoon's tracks and Billy and some of his 3rd platoon cohorts pulled up at a tank trail intersection. The wallet's owner came forward and greeted Billy. The greeting was special...the grateful A-trooper and Billy shook hands earnestly, as brothers might do. It was late in the afternoon and a dozen or so members from both Troops were on hand.

As Billy and the A-Trooper talked, we couldn't hear exactly what was said, but there was no doubt about how grateful the A-Trooper was to his fellow Guardsman from Virginia. It was an impressive moment, worthy of the round of applause that broke out right there on the tank trail.

An editorial note about A-Troop. Although A-Troop is part of the MTOE 223rd Squadron, it is conjunctively made up of the socially elite of the city of Philadelphia, They own their own armory, a small herd of ceremonial horses, some number of functioning military vehicles, are "voted" into membership, and they all donate their military pay and allowances to the social component of A-Troop, AKA The First Troop, Philadelphia City Cavalry. One member has published a book entitled The Gentlemen Of Gloucester. A-troop trains but when it comes to labor such as cleaning, maintenance, cooking, and such menial tasks, it is (or was) all hired out to a contractor who wears military equipment and relieves A-Troop to enjoy the schedule of the day. E.g., the rations issued are not consumed by A Troop. They give the rations away and their contractor chefs prepare a more gentlemanly fare for the members.

Joining the Guard in AR
Carroll Childers, MG, ret

The Soldier who became a Major General and Commanding General of the 29th Infantry Division on 23 Aug 1996 is the young man who celebrated his 17th birthday on 24 Aug 1955. The school cycle which began in September following this birthday brought into his life the opportunity to start his journey towards being a Commanding General. Sometime in the later part of September 1955, the Crossett High School Principal announced that the Senior Class boys would assemble in the basketball court grandstands for a meeting with First Sergeant Hershel Williams to hear a presentation on a National Guard unit which was being formed in the small town. We had no idea what to expect but we knew that if Mr. Willis told us what and where and when, we did not need to know why; yet.

When the First Sergeant was done, I had little idea of what he had said other than that joining the National Guard meant attending a 4 hour meeting once a week, going to what was then referred to as Summer Camp for two weeks a year in the summer, and that we would get paid for those 4 hour meetings and for that 2 week summer camp. He mentioned who the Company Commander was, Robert Downey, whom most of us knew, either personally or knew about him, because he as an important figure in the local Paper Mill.

The First Sergeant did not tell us much at all about what kind of National Guard unit he wanted us to join, i.e. he did not tell us what kind of unit it was and what it's mission was or what kinds of jobs were part of the organization. We didn't know enough to inquire.

This initial meeting was one week and sometime the next week would be 3 Oct 1955. All of we who were impressed with the initial meeting, reported to the local Fire Station for sign-up because the Guard had not finalized preparation of an abandoned school property which would act as an Armory for several years. All of us who signed up this night then were assembled for swearing in and other administrative processes.

"OK soldiers, the first thing we have to do is assign jobs to everyone."

A small dull roar arose as scores of small conversations grew out of that brief announcement.

"Awright now, cut the chatter. You are in the Army now. Keep your mouth shut until you are called on and then stand up, sound off with your name and rank and give the shortest answer you possibly can. If you are talking to me you will end the answer with FIRST SERGEANT" (and he screamed these two words so loud it hurt). "If the Capn is speaking, you end the answer with SIR. If Platoon Sergeant had asked a question you would end the answer with SERGEANT."

"QUESTIONS?" Some fool raised his hand.
"YOU. With you hand in the air. What do you want?"
"Sergeant, I have a question."
"I am the FIRST SERGEANT Soldier; it is too early to have a question and you are not following directions very well. Get out here. RIGHT NOW, get out here on the floor, front leaning rest position. Sergeant Ward, show this soldier how to assume the front leaning rest position."

Ward took the front leaning rest position in preparation to demonstrate the pushups he knew would follow.

"Are you watching Sergeant Ward soldier?"

Williams takes a few short back and forth steps and addresses the rest of the soldiers, clutching a clipboard to his chest with the left hand; right forearm parallel, fist clenched, index finger pointing down, forearm pivoting violently about his elbow in perfect synchronism with his articulation of each word.

"Is everybody watching this? Better be watching cause you will be doing lots of these."

Suddenly Williams comes to attention; "Demonstrate a push up cycle Sergeant Ward."

One, two, three, four. Recover.

"Now you see everybody, a push up is not just a pushup. It is two down and two up for a count of only one pushup. Now, soldier, what was your question?"

"Sergeant Williams, I do not know what my rank is; what is my rank?"
"You ain't got no rank Soldier. You are a Private E dash nothing."

Then he addressed the whole assembly; "All of you are. You are all E dash nothing. You are lower than anything I can think of right now. Take your seat Soldier."

"Awright now, where was I before I was so rudely interrupted? Jobs! Awright, who wants to be a clerk?" No response. He looks around slowly.

"Awright, we gotta have a clerk otherwise all the paperwork gets screwed up and nobody gets paid. Got to be able to type. Who can type?"

A couple of hands went up rather meekly, perhaps hoping not to be noticed. Two hands. Good. That is exactly how many clerks we have.

"Awright, next, who wants to be in charge of food preparation? Be working for Sergeant Billy Joe Sharp there. Work one day and be off one day," he drawled.

That had an appeal to me so I raised my hand. OK, here is one. What is your name soldier? You are now a First Cook. So on it went through a number of other jobs including mechanics, truck drivers, supply personnel, radiomen, mechanics, and I cannot recall all that was covered.

Then, out of the blue came the request for "Who wants to be a tank Driver"!?!?

"ME, I want to be a tank driver." I jumped up and waved both hands over my head.

"Set down and shut up you fool; you are a cook and you cannot be both. I need 17 drivers in all."

The First Sergeant continued to write names for tank drivers. I sat there feeling like I had been cheated and somewhat lost interest in the rest of the evening. I barely knew what they told us about the next step or the next drill.

Learning the trade of being a First Cook and operating and maintaining the kitchen equipment kept me occupied for the next several months of weekly drills and probably in July of 1956, The Company loaded up in Military Transport trucks and departed to drive from southern AR to Camp Polk LA. This was the longest drive ever in my life and doing it in the back of an Army Truck was no fun at all.

Upon arrival, I had had approximately 146 hours of on the job training in the preparation of food and service. We were preparing food in a hard building which was called a Mess Hall. Food was prepared on the mess hall equipment and all the auxiliary equipment which supported a Mess Hall. Perishable food was stored in large walk-in cold rooms. Fresh issue of foodstuffs was made each third day and it was up to we novices to manage that supply so that it ran out precisely every 3 days. The use of the word novice is important because in the 4 hour per week training routine we experienced, we never cooked a meal. So here we were, about to start a two week training cycle and the only member of the crew who had ever cooked and served military rations before was the single sergeant E-7 who wore the tall white paper headgear indicating the Mess Sergeant.

The way things were organized was that a mess hall was shared by two Armor Companies. Our Mess section would feed two companies one day and the mess section of the other Company would feed both Companies on day two and so on alternating through 2 weeks. This is how the mess Section achieved one day on and one day off.

It was my third summer at Camp Polk, LA and I had yet to drive a tank and I must admit that I was not a happy camper. I made up my mind that if the command would not, I would just force a change myself. A bout half way through serving the other Company on the first noon meal a really big guy is making his way down the line. I knew I could outrun this slob so as he stood in front of me and gawked at the very small amount of potatoes I served him, he complained "Hey knucklehead, you gave your guy here (nodding to his left) lots more potatoes than you gave me."

"You want more potatoes do you," I asked.

"Hell yes I want more potatoes," he growled. So I dipped deep and issued him an extra spoon from about his nose down to his chin. They were, shall we say, warm. I knew my next move and that was to dash quickly through a familiar path to the back door and disappear among the WWII era buildings.

I went straight to the First Sergeants office and turned myself in. Next morning I had been reassigned to a tank crew.

Whistlers
SPC Latham

My first driver was a fellow named (redacted). His brother (redacted) worked as an HVAC installer. He was working way up there in the roof (inside) of the Richmond Coliseum during construction. He told me he dropped a "whistler" from his perch high above, into the sand below and held on for dear life. Police, fire, Ft. Lee bomb squad.....they all showed up.

About a week later, he dropped one of those "whistle only" things and cleared the place again.

Legends, Rumors and Myths
LT John Terry

When I enlisted in C Troop in Jan. of 1969 I had no idea what kind of unit I was joining. However after a short time I realized that I was in a very special unit and some exceptional men. Whether it is the 7th Cav at the Little Big Horn or the Charge of the Light Brigade, the Calvary holds a colorful image in history. And C Troop is no different. While C Troop contributed to the history, we have to give credit to the various troopers that came before my introduction into the unit in early 69.

The following are short stories that were passed on by senior members of the unit, some of these troopers go back to the transition from the Light Infantry Blues to the 183rd Cav Sqd in the 60's. Most were told at the C Troop Club, a clandestine organization that did not exist on paper but was in fact a real and important part of the unit. Troopers such as Johnny Thomson and Ducky Sullivan were instrumental in keeping the oral tradition alive.

Throughout the state the unit had both a famous and infamous reputation. The first story I remember was a Lt who lost his finger. Seems that the newly organized 183rd Cav with units in Richmond and Fredericksburg were assigned the M 41 tank. This tank had blank rounds to use during field training. After a long day training in the heat of the summer the unit was on the wash rack hosing the tanks down. At this point I do not know which unit this was, but it could have been the troop from Fredericksburg. The Lt being a real Cav officer was helping to wash the tank when he put the hose in the barrel of the main gun and it exploded. The explosion ripped his finger off! After immediate first aid and transportation to the appropriate medical facility, the crew went back to cleaning the tank. When one of the crew members dropped the breach inside the tank, there in the breach was the Lt's finger.

Apparently the tank still had a blank round in the breach and the heat and cold water combination caused a cook off, thus injuring the Lt. This represents one more event in the growing legend that would become "C Troop."

For the most part, there is much truth in the story about a crewman losing digits in an incident with the M-41 tank, but I believe the correct story is as follows.

My first Annual Training (AT) after commissioning was at A.P. Hill VA, the summer of 1965. I was a Platoon Leader of C Troop, 1-183rd Cavalry Squadron, 29th Infantry Division. Captain Alvin York Band was my C Troop commander and Archibald Sproul was the Division Commander; a D-Day veteran commanding the HHC that fateful day. He was awarded the DSC for Gallantry.

We had been training hard all week and as I recall it, the accident occurred on a weekend. We had just completed a demanding tactical exercise, had topped off with fuel (the M-41 tank was a gasoline burner) an were directed to move to the next hill mass, back into a Troop Defensive position in the edge of a wooded area, hull defilade positioned, and assemble in the center of the horseshoe for an After Action Review. Several things had already occurred without the knowledge of the crew which set us up for the accident to follow.

First, it was a very hot day; 106 degrees. When the fuel tank was topped off to specification, the nominal temperature and what that would do to the level of fuel in the tank was not calculated. Second, and un-noticed to anyone, the fuel cap gasket had broken and was absent. Add one and two, gasoline volume expanding against a unsealed gas cap, then, when the driver backed into his hull defilade position and hit the brakes causing the chassis to rock considerably. Raw fuel poured out through the unsealed cap and on to the hot engine. Also, M-41s had a bad habit of backfiring when shut down so this may have added to the evolving accident. What it sounded like was backfire/kaboom and the entire rear of the tank looked more like a fighter jet in full afterburner than a tank.

All crewman got out of the tank without injury but for the loader. He was reaching back into his hatch to recover some of his gear and the flame licked up and removed all of his eyelashes and eyebrows.

My command jeep was right adjacent to the tank and I radioed the situation to the C Troop commander who called the post fire department. In about a half hour the first fire engine arrived, reeled out its hose and activated the pumping system to fight the fire. The engine did not have a drop of water in the tank. Ten minutes more and a second engine arrived. Same story; no water. The tank is burning fiercely. There is a lot to burn on a tank and of course various pyrotechnics had been going off steadily for several minutes. Finally, a engine, with water, arrived and began to extinguish the flaming tank.

Now comes the real accident. After spraying the source of the fire somewhat, a fireman, for some unknown reason, made the decision to cool down the main gun tube even though he had been cautioned that there was a blank round in the breech that had not cooked off yet. At the precise instant that the fireman stuck the brass hose nozzle into the muzzle, the blank cooked off and launched the brass nozzle like a bullet.

I was under the impression that he lost 3 fingers in the incident. They were discovered later in the tank floor. Many estimate that the fingers impacted the muzzle flash suppressor, remained there briefly, then basically slid down the tube and fell out the open breech. The soldier who discovered them, damaged well beyond reattachment, first thought he had discovered some Vienna sausage but with further examination realized these had once been fingers.

As the Platoon Leader, I got the opportunity to give Major General Sproul an account of what happened as I knew it.

Another story is about how one of the troops totaled a tank. Two tanks were hauling roaring the tank trail in the dead of summer. The trail would have so much dust that it would it be very difficult to see if you were in trail. Seems these two were very close and the first for whatever reason swerved to one side and the second tank was unable to react in time and hit the first tank on an angle on the tank track. The lead tank had its final drive sheared off. Since this could not be repaired the tank had to be totaled. Everything and I mean everything had to be pulled off the tank. Main gun, power pack, vision blocks, range finder, radios, computer, even the hatches and seats. Anything that could be used on another tank. The hull was towed to the tank gunner range and used for target practice. Again one more event in the growing legend of "C Troop."

For many years C Troop was banned from going to SMR to train. That was reversed by Capt. Childers in the mid 70's. I was told (at the club) that the reason the Cav was prohibited from Camp Pendleton had to do with an incident that had occurred during the late 60's. It seems that in those days units would try to schedule training at the facility in the August to September time period in order to use the beach for an after AT celebration. Which Cav Troop is responsible has been lost to history but the story goes that this unit did drill and had their party on the beach that afternoon and evening. The next day as was customary the men cleaned the barracks and lined their jeeps and trucks up for convoy back to the home station. One driver could not find his jeep. After frantically asking and searching one of his fellow troopers recalled that the driver drove it up to the beach the previous evening. Hitching a ride up the beach, he found his jeep. It was about 20 yards out in the surf with only the antenna swaying in the water as the waves came in and out. Not sure how much damage this caused but I'm sure the paper work was a bear.

Which brings me to another revelation. The AST's (administrative supply technicians) for the Cav were very special. I cannot comment on the period before 1969, but during my stay in the unit these full timers had an extra amount of work from all of the mishaps that were a natural part of hard training with large and dangerous equipment and armament. While members of the 223rd Cav, C Troop did have a HHC Troop in Philadelphia, but due to distance I can't believe they were much help. When C Troop was reorganized to 183rd and assigned as part of the Brigade base they were without the squadron staff to support.

One example of extra work involves the destruction of a 4.2 in. mortar. We were firing illumination for the tanks, when a round hung. After going through the immediate action and waiting for the barrel to cool down, it was now time to pick the barrel up and turn it upside down and catch the round as it falls out of the tube. I was the lucky guy that got to catch the round. In theory that should be no problem, since it takes 40 revolutions to arm the round. Sounds good but in theory it is not supposed to hang either. Well it would not fall. No matter how much shaking. By the way it is not easy to shake a 4.2 in mortar barrel! We notified Range Control and had two troopers spend the night with the tracks at the firing point. Range Control contacted Ft. Lee's ordinance section and they came out the next morning. After hitting the barrel with a sledge hammer which I heard almost gave a few troopers a heart attack, the ordinance team took the barrel out about 300 yards, strapped C 4 on the tube and blew it up. Can you imagine how much extra paper work that cost our ASTs. They were exceptional! But again this added to the reputation of C Troop.

Little yellow tanks on the side of helicopters! I'm not sure where or when our competition with the fly boys started. Could it have been with Delta Troop of the 223rd? This particular event occurred at Ft. Picket, during AT around 1975. When C Troop was not night training the men would sometimes get into trouble. One night, a small group (mostly scouts) decided to paint yellow tanks on the sides of the helicopters which I think belonged to the 76 Gun Bn. After that the Brigade MP's bet C Troop that we couldn't paint those helicopters assigned to HQ. That night a small dedicated group crawled on their bellies and accomplished the mission. Apparently the MPs thought that by doubling the gate guard you would deter the Cav. They under estimated the Cav badly. The next afternoon the MPs arrived with a case of beer that was immediately donated to the C Trooplub to be enjoyed by those that liked an adult beverage.

However this is not the whole story. Only now, after the stature of limitations has expired, can the rest of the story be told. While some men were out crawling across the air field, Capt. Childers and Lt Deverell had taken a jeep and were out riding around. A past member of C Troop came by the BOQ looking for some trouble to get into. Lt Redfern and I being young single guys needed our beauty sleep, we chose not to participate. The Capt. and Buddy had been stopped by the MPs and were detained because they didn't have the log book to the jeep. It took an hour or two to prove that the jeep really belonged to the Capt. and was on his property book.

The past member decided to act on his own. He went upstairs to one of the pilot's rooms. He soaked a roll of toilet paper in water and inserted a grenade simulator in the center. Pulling the string he threw it into the pilot's room. The explosion covered the room with small wet pieces of paper. Like a thousand spit balls!

The pilot upon returning to his room took it all in stride. The next morning he wrote a note of apology for the condition of the room and left some money as a tip for the maids extra work. If only that were the end of the story. Turns out the culprit also attached a trap to the waste basket. When the maid tried to empty the basket it went off.

First it went up hill to the Post Commander and then started downhill to the Brigade Commander (I think it was BG Bradshaw) who called in the commander of the helicopter unit and Capt. Childers. He wanted some answers! But the Captain and XO were detained by the MPs, therefore they had no knowledge of this incident. Another part of the legend that we call C Troop.

Not all of the legends and myths and rumors have to do with equipment or mishaps. There is the reputation of A Troop of the 223rd. This unit like C Troop can trace their lineage to the mid-17th century. The rumor is that they were very selective in who they enlisted. Handpicked their officers and gave back to their C Troop Club all of the pay for UAT's and AT. Now the legend goes that the annual training at Ft. Drum the unit rented a large boat and sailed around the Thousand Lakes area of Canada the middle weekend of AT. Their First Sgt had hired substitutes for any duties needed over the weekend.

While I cannot verify that, I can tell you about another time C Troop came down to Ft. Pickett for AT. They are rumored to have procured a motel on the beach and hired a band for Saturday night and gave free invitations to all of the attractive girls on the beach for the party that night. I wasn't there but I was across the road when they arrived at their barracks. There were three U-Haul vans. They started unloading, one with cases of beer, one with what appeared to cases of whiskey, vodka and gin, and the other with cases of sodas. And the legend continued!

Lastly there is the story about the battle field in Fredericksburg. It seems that the locals and tourists were getting extremely disturbed! A lone tall jogger in shorts and tee shirt with protective mask and hood was seen frequently running through the park. Apparently this National Guard member lived in the area. The rumor goes on that the Park Service asked him to stop jogging in the area and he agreed. To this day that jogger has not been identified.

These are but a few on my memories.

Sgt. Lindsay Bruce, the "M60 Bandit"
SSG Roddy Davoud

Summer Camp of 1982, at Fort Pickett as best I can recall. Early in the first week, our scout section was in the field doing recon. Lindsay Bruce and I discovered a hidden defensive position of an "enemy" force from another unit. Lindsey and I snuck up on the poor young soldier, manning the position alone, by using Lindsay's Marine training techniques. He called it "crawling through the kai-kai weeds." We "captured the enemy" who was practically scared to death. Lindsay took the young soldier's M-60 machine gun as a war game trophy after threatening him with all kinds of physical damage if he ratted us out and off we went.

Apparently, no one could figure out who the M-60 bandits were so they didn't know where to look. We sure didn't confess to having the weapon, but a couple of days later the stuff hit the proverbial fan. Word came down from on high that everyone was going to be forced to stay on base middle weekend if the missing weapon didn't somehow "re-appear." I was the senior NCO so I accepted the responsibility of telling our Platoon Sergeant that I just might know where the weapon might be found.

Not another word was said and we all went home for the weekend.

Meeting SFC Macon (Mercury) Morris
SSG Bob Anderson

Macon Morris was platoon sergeant and tank commander in the first platoon in the 1970 - 1975 and later, era. He'd been around a while, probably about 20 years older than many of the newer troopers just joining up. (At this time, Ducky Sullivan was second platoon sergeant and Johnny Thompson was top NCO in the third platoon.)

Macon was not a big fellow. "Vertically challenged," Morris stood about five-six in a new pair of boots and soaking wet, may have tipped the scales at a buck fifty. A razor-clean Flat Top was always his hair style of choice. Setting the example at drills, he always looked like he had just climbed out of the barber's chair. For those old enough to remember the low-key, dead-pan delivery of comedian George Gobel (aka "Lonesome George"), Macon Morris was his look-alike, and act-alike.

SFC Morris was calm, cool and collected. His slow drawl gave away his southern roots. Macon was definitely not a New Jersey American. His manner of speaking was matched by his deliberate, ambling walk. When he needed to look at something, his head didn't turn, he turned (slowly) his whole upper torso toward whatever drew his attention. In conversation, he would listen intently, then pause while he conjured up a proper response, usually an in-kind, barbed retort. Folks learned fast in dealing with Macon that his slow drawl was just a disguise for his quick wit.

One MUTA-5 at Pickett, Morris showed up at Tank Range Five driving a brand new Japanese (an Isuzu or Mitsubishi, perhaps) pick-up truck. Back then, these small import trucks were just being introduced in the states. Ford's F-150 was literally twice the size of Morris's new ride. And, the color he picked was odd, sort of a blend of mustard yellow and English pea green. He was obviously proud of his new acquisition. But, of course, the gathered throng enjoyed ribbing him about the tiny truck, saying things like (while bowing) ..."Ah-so, little honorable pick-up truck"..."Can you get a sheet of plywood in that thing?"..."Motorized wheelbarrow," etc. Morris loved it...he stood there grinning ear-to-ear while his friends ribbed him about that truck.

Macon got the nick-name "Mercury" after the famous running back Mercury Morris, a super star for the Miami Dolphins. Macon's whole persona was slow, deliberate — the exact opposite of the super quick, shifty Miami Dolphins' runner. (The Dolphins were quite the team in the early 1970s, under Coach Don Shula, winning Super Bowls with star players like Bob Greise, Larry Csonka, and Mercury Morris.)

R. T. (Bob) Anderson, joined C Troop as a transfer from a Hampton-based Army Reserve outfit that consisted of parachute riggers. Part of rigger training at Fort Lee, included Bob packing his own chute and using it to jump out of a perfectly good airplane. Talk about paying attention in class!

Coming to C Troop, Bob knew nothing about tanks much less how to drive one. He learned how to a drive an M-48, A-1 "on the job," driving first for Joe Keesling, a first platoon TC. Then, when Macon Morris needed a driver, Bob got the gig. Bob and Macon teamed up for the first time at Camp Pickett in July, 1971 for Annual Training. Daily high temperature at Fort Lee in July is 95 degrees with relative humidity to match.

Anderson, a quick study, caught on instantly as to how Morris wanted his tank driven. Morris's first instructions came over the intercom as they left the tank park for the field on day one:

"Son, you can drive this tank as fast as you want...as hard as you want...as rough as you want but, when it stops, it better be in the shade."

Roger that!

My Interactions with TAGs
Carroll Childers, MG, ret

The first Adjutant General I remember was a feisty little short guy named Paul Booth, Major General by rank, Adjutant General of Virginia by position. Behind his back he was referred to as Small Paul by such people as our First Sergeant Jiggs Brennan and several C Troop commanders in the Cavalry Squadron. Nobody would ever call him that to his face. His wife was a good head taller than he and seemed to always have a very large "Southern Bell" kind of a broad brimmed hat on. She smiled a lot and he always had a sour look on his face like he would rather be somewhere else doing something else.

My one and only close encounter with Booth was when I was a candidate in OCS at Virginia Beach. He was such a non-descript, non-military looking kind of guy that one might just think he could be the yard man or a PFC. He wore his headgear sort of cocked off center and ambled along rather than moving like he had a mission. I was hurrying from point A to Point B one day between buildings and he was walking by and I just did not recognize this little guy at all. I got about 3 strides beyond him and someone yelled out extremely loud so that I knew that I must be the intended recipient for whatever it was that was obviously about to be dished out. I stopped immediately in mid stride, came to attention, did an about face to assess the situation, and there were two TAC Officers standing by The Adjutant General (TAG). Where they came from, I don't know, but they were enraged that I had ran past TAG without acknowledging him. I should have recognized him, slowed to a walk, saluted and greeted him, and continued my run after he returned my salute and gave me a carry-on command. The two TAC Officers worked me over for 5 minutes or so while Booth looked on approvingly, gave me a gazillion pushups and squat-thrusts, and threatened me with the most god-awful punishment if I ever again failed to show appropriate recognition to seniors.

I did encounter Small Paul once more in my career; at my commissioning exercise the following summer. I was the Distinguished Honor Graduate (DHG) of the Class so I had a personal encounter with TAG as he gave me some of the prizes accorded to the DHG. I always wondered if he ever made the connection between the Candidate who passed him by and the most recent DHG. Because TAG is a political appointment in Virginia, Booth soon lost favor with new Governors and was succeeded by then LTC William J. McCaddin.

McCaddin had been a Major when I first interviewed for OCS. An impressive and impeccably dressed officer, he was a veteran of Korea and was just obviously destined to move into the stratosphere of the National Guard. He was smooth, smart, cool, calculating, articulate, and had an air of sophistication that seemed to go along with being a very senior officer. Of course at that time, I knew little about anyone above a Captain.

The expectation had always been that the DHG of each class would accept the offer of a position as a TAC officer within the OCS. That was not my plan. When I got the call to tell me that the intent was to have me assigned there, I called up my C Troop commander, Captain Alvin Bandy, and asked him what the deal was; did I have to do that?

"LT, what did you join the Guard for?" he asked.
"To be a Platoon Leader of Armored Cavalry," I replied.

"Then that's what you tell them. Tell them that you were commissioned in Armor and you have an assignment in the Cavalry Troop in Fredericksburg and that I want you as a Platoon Leader," he replied.

So I did. That was probably the first big mistake I made in my new career as a military officer. I went against what the system wanted me to do. Later I found out that McCaddin was furious. He was then the deputy Commandant, soon to be the Commandant and he took it very personal that I would turn down a slot in the OCS. But, I really didn't have much interaction with someone at his level so I more or less ignored the gnashing of teeth over there and pursued becoming the best small unit leader that I could.

As such, I soon gained a reputation for being a really good tactics instructor and the OCS began to employ me during weekends and Annual Training teaching tactics and running exercises. I was not a TAC Officer; they were the ones that screamed and hollered in the faces of the candidates and square them away in the barracks and on the parade field and herd them from point to point. I actually did not know what TAC was an acronym for; Terror, Anguish, and Chaos maybe? But oddly enough, TAC Officers did not teach tactics as the uninitiated might presume. The school employed others for this. So in addition to my regular drill schedule with the Cavalry Troop, I did additional duty teaching the OCS both on weekends and during the two week Annual Training period. By the time I stopped teaching tactics, I had taught 19 different Classes. I did not teach every year as I sometimes had conflicts in schedules.

Time flew by with several National Guard reorganizations and two Civilian deployments for me to Vietnam as a civilian representative to the Vietnam Laboratory Assistance Program. These two 4-month tours, one with the Mobile Riverine Forces and one with the USMC, added greatly to my reputation and to my skills and experience in a real combat environment. I dealt a lot with Navy SEALs and with Marine Force Recon so brought a lot of interesting tactics and techniques back to integrate into the training. I had made it to Command of C Troop 223rd Cavalry in Richmond. We were part of a Squadron headquartered in Pennsylvania but most often responded to and trained with an infantry brigade in Virginia.

The 116th Infantry Brigade was commanded by COL Julian Bradshaw, a genuine infantryman who had landed with the 29th Infantry Division on Omaha Beach, 6 June 1944. I never knew why COL Bradshaw didn't like me but there was always that little crevasse between us. I think I had some sort of reputation that he either did not approve of or he was envious of or some combination of these. Certainly, the Cavalry Troop was always doing unconventional acts in training and sometimes wore clothing and accoutrements that were not necessarily approved (Cavalry hats, bandanas, camouflage clothing, etc). By now, including having served in C Troop 183rd Cavalry under Bandy, I had been a Cavalryman for 8 years or so and "we" had gotten a lot of attention with our training and antics. Bradshaw didn't have the personality to simply call me on the carpet so he just harbored his pent-up emotions. This is all in hindsight on my part, and perhaps a great deal of supposition.

A little set up is required to continue this tale. The time is now early 1973. The Vietnam War is over but for the formalities. The draft has been ended. Recruiting and Retention (R&R) is really tough. There is no other way to say it. Guardsmen were viewed as draft dodgers, then the Kent State shooting of students by Guardsmen did not help the overall image of the Guard at all. Little help is offered by the state and there were no enticements of any kind to attract new recruits as is the case today (2018). Long hair was in vogue on the street. I had a workshop involving the whole Troop one Sunday afternoon to talk about how we could meet the R&R goals. Lots of good ideas to try; some I felt good about and some I had some reservations about. Unlike their counterparts on active duty, commanders in the Guard were responsible to keep the ranks filled. A long list of things attempted could be presented but there seemed to be little that really worked and maintaining strength was a consistent, continued problem in need of a solution; so approaches were in constant motion and I was continually trying to find mini-solutions. A group of NCOs advised me that retention in the C Troop would be enhanced some, thus those retained might be inclined to recruit their buddies, if I would relax my standard on individual grooming; haircuts and facial hair. Guys in this period were attracted to wearing their hair long and to growing facial hair. Moustache was a particular challenge because the Army standard was very precise versus the style that soldiers wanted to sport. It was not that the soldiers wanted to go to extremes, but the style they wanted to go to just didn't fit.

My view was that if I allowed an infringement for one rule then I was off base on enforcing standards on other things. It was a matter of enforcing all regulations or throwing the book away and I resisted that perhaps to the extreme. One I had reservations about was hair. HAIR!! Black soldiers were wanting to do things with their hair which did not look military at all. White soldiers were into pony tails and sideburns and facial hair. I agreed that short hair wigs could be worn on duty and I gave up some scrutiny in moustache inspection....but only for one year. If R&R saw benefit, it would continue and if not, out with the wigs.

This would be a one year trial period. Everyone knew that if strength did not improve in one year, it was back to more scrutiny and enforcement.

William J McCaddin had been appointed as the Adjutant General of Virginia. It crosses my mind that he could just be holding a grudge for our impasse in that OCS assignment I refused. Break for explanation here. There is another CAV Tale that mentions a grudge against me but that story did not involve MG McCaddin.

The Brigade deployed to Ft Bragg to conduct a 10 day long brigade level staff exercise; the biggest show in town for Brigades. I participated with some of my Platoon Leaders and NCOs. We assisted in setting up the entire event; raising tents, handing camo, running wire, setting up maps and work areas; two days to set up. I am out with pole climbers strapped on, scaling trees to hang camo above the tent rather than have it drape on and assume the profile of the tent. The Active Army flew over us and could not spot us. Once the map-based war games began, my folks functioned as the "eyes and ears" of the Brigade and conducted various map-based simulated operations in support of the Brigade's mission. The Troop is really making an impression on the Active Army observers not only with the map maneuvers and communications but with the advice provided to Brigade on prudent decisions. People are talking about Captain Childers from climbing trees to war game coup-de-grace. In all ways, the Troop is looking good and I am actually overshadowing the Brigade S-3, a LTC who was way over his head in tactics and command and control. I think if I had proclaimed a coup I could have taken command of the Brigade and everyone would have stepped back and let me call the shots.

Selected elements of C Troop were present acting as the CAV element in the command post exercise and, frankly, doing most of the work in physically setting up everything for the exercise. COL Julian Bradshaw is the Brigade Commander. Julian is a WWII Infantryman; landed on Omaha on about the 3rd wave and survived the war, decorated and wearing the CIB, a rare decoration in the National Guard in the 70's. In civilian life, he was a rural mailman.

So here he is, a Colonel, commanding some 5000 soldiers as a Guardsman and in daily life, he fills rural mail boxes; he is not even a postmaster. Quite a contrast in stature and I think it affected his ego somewhat. He tried to be all business, stiffening his chin and clicking his dentures sometimes as he gave guidance and acted as if he really knew what he was doing. Anyone who took any kind of exception or offered some suggestion or sounded as if they might doubt anything he had to say was under immediate appointment to discuss it with him later. Of course, that would be on more than one occasion. Later he wrote me a letter to tell me how much he appreciated my obvious study of the principles of war and my contributions to the command post exercise (CPX) being evaluated by the active army.

We had been at Bragg since Wednesday for a couple days setup, then the games began. My Cavalry Troop happened to be drilling this weekend. I had told the Troop via my standard Troop information letter the previous week that the year was up and that the wigs and bindings were no longer allowed. So those who were affected by this decision had about a week to begin complaining and learning how to file IGs. One of the soldiers mailed a copy of the letter to Major General McCaddin, then TAG VA. It was Saturday morning late and things were really going well in the CPX. Then, on the last day of the staff exercise, we had the enemy in deep kimchi and were destroying them with great efficiency. COL Bradshaw came in and asked me to go with him to his tent. He was red in the face with hands visibly trembling as he passed me a telefax copy of a letter and asked me if I had written it. Obviously I had. It was the Troop letter that I put out to formally end the hair policy.

"Yes Sir, this is my signature, this is my letterhead, this is the message that I sent to my troops AT MY PERSONAL EXPENSE. Yes sir. What is the problem?"

"The problem, Captain Childers, is that the Adjutant General sent this to me to confirm that you sent it. He doesn't like it. He wants me to fire you and I can see no reason not to."

"Sir, I tried an experiment. The troops knew the rules. The experiment did not work. We are going back to the original rules."

"That may be but some of your Troops have filed an IG complaint and that creates other problems. We don't have time to deal with commanders who generate IGs. You are too hard on the troops anyway so we would be better off without you. The CPX ends at 1800. You can leave then for Richmond and say your goodbyes to C Troop Sunday. I will expect your resignation on my desk sometime next week."

"Sir, my resignation may someday show up on your headstone but it will never be on your desk." And I saluted, did an about face and left him standing there.

I found a phone booth and made a phone call to Alvin Bandy. He was no longer in the Guard but he was my career mentor and he knew people. One of the people he knew well was COL John G. Castles. Castles was a unique officer. He was a combat infantryman in WWII and highly regarded. I had encountered him in tactical play in the field when he was a Battalion Commander and I a Platoon Leader. Later, he sought the appointment as TAG in competition with McCaddin. Castles had lost a leg from the knee down in an accident after the War so when he began a campaign to get congressional support for his appointment, Bandy enlisted me to act as his aide driving him around Washington, carrying briefs, and generally keeping him company. So I had gotten to know him as a junior officer but more importantly, he had gotten to know me. Yet, I would not ask him to intervene for me with McCaddin. I knew that McCaddin still held the OCS event against me. He was small that way. So I told Bandy what had happened and asked him for advice.

"Look", he offered, "don't worry about this. Bradshaw is a hillbilly. McCaddin may or may not have given him specific instructions. Let me talk to COL Castles about this and I'll get back to you."

By the time I got to Richmond, one of my key mentors of my career, who would intersect my life numerous times, had been briefed by Bandy. John G. Castles at the time was a Colonel and the Chief of Staff to McCaddin. Castles was a WWII Infantryman, decorated with Bronze Star and CIB, and a most highly respected combat leader. In my earlier years, as a cavalryman in C Troop 183rd out of Fredericksburg, Castles had been an infantry battalion commander, LTC, with whom I had sparred against in the field. Also Castles and Captain Bandy were great friends. Both were each on the county board of supervisors in their respective counties and knew each other there, in the Guard, and in social circles. Castles was also quite wealthy, independent, and insulated from any sort of intimidation. He commanded an infantry battalion wearing a prosthetic. Here was a no-BS leader.

Castles got involved with this incident and from what I was told much later he went to McCaddin and berated him for caving in to the prospect of some minor IG looking into a good commander trying to do his job. "There is no IG case here. Let em dig. There is not a better company grade officer in the state than Captain Childers and you want to can him because two enlisted soldiers disagree with his decision as a commander?!?!"

The entire incident simply melted and I never heard another word. But it did add another black mark to my ledger. It also revealed that I was viewed very favorably by a key senior officer. This fact caused yet others to build their own personal case against me as they were calculating their own personal rise in rank and now there is this renegade, this Cavalry officer that is outside the state norm (infantry, artillery, intelligence, logistics) may have some leverage and somehow pass them by. So Majors and LTC's began to view me as a threat. It was one way, in fact, it was many years later before I even gave such incidents and maneuvering any hindsight analysis to conclude what I offer above.

After some 5 years as C Troop commander I was reassigned to the 116th Regimental HQ and stayed there for over two years. The first Annual Training period as the Asst S-2 I established a classical enemy command post (ECP) with protective Ops and active patrols to protect it. The various Battalions were given 1 day missions to send reconnaissance teams out to find the ECP and to gather as much information as possible without having an encounter. It was a most productive exercise and became quite competitive and a grand success that was talked about for several years. The next year I was promoted to S-2 and was involved in staff training for most of the year. Then, COL Bradshaw retired and COL Castles became the Regimental Commander.

A few years later, The Honorable John O. Marsh (Congressman VA) became the Secretary of the Army. Marsh and Castles were fellow soldiers and the best of personal friends. Castles was appointed as the Adjutant General for Virginia and served in that position for 12 years.

He served under both Democratic and Republican Governors and only retired after losing his Federal Recognition at age 65. Some TAGs have been known to continue to hold the office after loss of Federal Recognition, but Castles was a man of great principles and would not do that so he stepped down at the first political opportunity, meaning at the next Governor's election so that the incoming Governor would have the opportunity to select a TAG of his own choosing. But to his last day in office, he remained my key mentor and benefactor. General Castles passed away in 2005 after a brief hospitalization.

To this day, I visit his gravesite regularly just to stand for a moment and honor him.

By the time General Castles stepped down from TAG, I was a General Officer. Getting to be a General Officer among the senior officer corps in Virginia was a real challenge. There being only a limited number of GO slots in the state, it was a cut-throat climate at every step. I had no real plan to become a General Officer. As I mentioned earlier, I was always happy to be where ever I was assigned.

Let me talk a bit about this cut-throat environment and those that inhabited it. General Castles took me out of Battalion Command in 1984 so that I could attend the Army War College. It would never have occurred to me because I was excited about being a Battalion Commander leading 556 soldiers. But Castles was looking ahead; he knew things that I did not know. For example, because of his relationship with Jack Marsh, he probably knew that the 29th Infantry Division was going to be stood up (brought out of retirement where it had been since 1968). Jack was a man immersed in history and tradition and simply had a personal goal of seeing the 29th Infantry Division back in the lineup. The 29th was stood up as a Light Infantry Division of exactly 10,000 soldiers based on, I think, two factors: the ability to transport an entire Division by air and some cost constraint associated with the Office of the Secretary of Army discretionary spending.

The Division stood up in October 1985. I graduated from Army War College in June 1985. I think Castles had in his mind that I would be a Brigade Commander in the new Division and work up to one then two star and command the Division; at least, I think he was posturing me for that potential. Unfortunately, an officer who held the grudge against me for the Flame Thrower trick held a great deal of influence in the Virginia National Guard and used his influence to lobby against my appointment. So I was excluded from being in the Division until this officer retired.

Meanwhile, I was given command of a so called "Troop Command." This unit was located in the Richmond Dove Street Armory where I had spent many pleasant years in Cavalry. It was composed of all of the odd units that were required to accomplish the state mission and provide support of various combat and combat support/combat service support units, but were not TOE to those kind of units. I had a Data processing unit, a Band, a maintenance unit, an Aviation unit, medical units, administration units, and a few others. These units had always been rather loosely organized, trained, supervised, led, challenged, and developed. They were all individuals. They never went anywhere together. They never went to the field. They didn't know anything about being soldiers other than the few who retained some skills from previous assignments. This would be another entire vignette but suffice to say that after I discovered what I had, and what I did not have, and this took me two weekend drills to conclude through visits; the Troop Command went to the field for a weekend drill. A first for the Troop Command.

Thereafter, the unit began to emerge as a real command. People were becoming excited about coming to drill. Rumors were circulating down town in the state headquarters that good things were happening and of course such good news is bad news for those who might be hoping that I would not succeed at this command. But it was a command and I intended to make it shine like nobody would think possible. Three years went by very quickly and those soldiers were equal to any unit in the state by any measure of interest.

NEVER Forget Your Friends
Carroll Childers, MG, ret

A newlywed young man was sitting on the porch on a humid day, sipping ice tea with his Father. As he talked about adult life, marriage, responsibilities, and obligations, the Father thoughtfully stirred the ice cubes in his glass and cast a clear, sober look on his son's face.

"Never forget your friends," he advised, "they will become more important as you get older. "Regardless of how much you love your family and the children you happen to have, you will always need friends. Remember to go out with them occasionally (if possible), but keep in contact with them somehow.

"What strange advice!" thought the young man. "I just entered the married world, I am an adult and surely my wife and the family that we will start will be everything I need to make sense of my life."

Yet, he obeyed his Father; kept in touch with his friends and annually increased their number. Over the years, he became aware that his Father knew what he was talking about.

Inasmuch as time and nature carry out their designs and mysteries on a person, friends are the bulwarks of our life.

After many years of life, here is what he (and you) will have learned:

Time passes. Life goes on. The distance separates. Children grow up. Children cease to be children and become independent. And to the parents, it breaks their heart but the children are separated of the parents.

Jobs come and go. Illusions, desires, attraction, weakens. People do not do what they should do. The heart breaks. The parents die. Colleagues forget the favors. The races are over. But, true friends are always there, no matter how long or how many miles away. A friend is never more distant than the reach of a need, intervening in your favor, waiting for you with open arms or in some way blessing your life.

When we started this adventure called LIFE, we did not know of the incredible joys or sorrows that were ahead. We did not know how much we would need from each other.
Love your parents, take care of your children, but keep a group of good friends. Dialog with them but do not impose your criteria. The National Guard develops friends like no other experience I am aware of.

Officer Candidate School (OCS)
Carroll Childers, MG, ret

Nothing can be compared to immediate family; spouse, children, parents, siblings. So when one says "the best thing that ever happened to me was" (fill in the blank); this is a statement exclusive of the immediate family and inclusive of all other categories of events that one might discuss.

In the inclusive category, many events have intersected my life. I think I am one of the most fortunate people that has ever lived as I have had the good fortune to visit 27 different countries, meet some of the world's most famous people, make contributions to mankind, save precious lives through engineering ingenuity, live a relatively uncluttered life with only a few health complications, be assigned to revered positions and posts, be a part of history on more than one occasion, be an example for others to follow, be awarded a wall full of certifications and accomplishments, and gain considerable respect and honor personally and professionally in both my civilian and my National Guard career. And I did this all on one wife. Much of what I have done is because my wife has always been so completely understanding and supportive of whatever I wanted or needed to do. Unfortunately Dayle, my soulmate for 56 years, passed away 14 NOV 2014 after battling cancer for 10 years.

Among all of these, two events stand out as those most important and rewarding to me. Because of these two, many of those intersections of my life have been possible. Officer Candidate School is one of the two most important events of intersection in my life. It postured me for so much of the remaining intersections, in fact, if not for OCS, the second intersection would probably not have occurred. So perhaps I can say that OCS has been the pivotal event in my life. Completion of Ranger School is the second event in my life which provided much leverage to my eventual achievement of Division Command. This clearly set me apart from any peers because not only was I awarded the Ranger Tab but I was (and still remain) the oldest Distinguished Honor Graduate in the history of the course (42 years, 8 months). Much of the secret to achieving this was the lessons learned in OCS; that much of the "harassment model" of the school is simply a tool for applying pressure to sort the determined from the rest.

Little did I realize that the simple commission ceremony at Virginia Beach VA in August of 1964 would set me on my life's path that would culminate as the Commanding General of the Famous 29TH Infantry Division. Once committed to the responsibility of leading soldiers and setting the example of those many qualities that separate the ordinary citizen from the patriot, I was on a path that I simply could not abandon. This duty to not abandon was somehow fermented in the OCS environment. There was deep sense of teamwork and dependence learned there that could not simply be forsaken just because OCS was behind me. OCS was some sort of conduit or conductor straight back to the Revolution and the sacrifices that so many people had made before me, which allowed me to attend OCS and I felt compelled to not short circuit that flow of energy but to pass it to every soldier that I could connect with.

I became addicted to service and to helping others, appreciating the need for service to the country. OCS was some form of kick-start; an enthusiasm pill; a trial that I was extremely proud to have passed; and it gave me a very special commission - The commission to be a leader of warriors. I thought there was no greater honor than to have the trust and confidence of our nation to train and lead warriors in whatever crusade might arise to confront us. Though I never got to actually lead warriors in combat, so many that I led, trained, inspired, and set the example for did go on to enter combat in a variety of assignments and they exceled.

It has been said that "They also serve, who only stand and wait." Certainly, this was the model for National Guard service during the Post WWII era, up to the 911 event. It was in this era that I enjoyed the 44 years of service to the nation for which I take great pride.

Many people that I have encountered came from a different commissioning scheme; academies, active duty OCS, and ROTC. Sometimes they attempted to demean a simple state OCS commissioning source as though it was secondary to their own. I took a peculiar pleasure in causing them to realize that they had misjudged my OCS. I never met a contemporary that I could not best because I had learned in the crucible of Virginia's OCS that when the going gets tough, the tough get going.

The environment of OCS challenges candidates in ways that they cannot contemplate immediately but these challenges will make them strong and resolute in so many and varied situations. They learn to overcome and adapt and accommodate. They learn to best the competition and not feel the need to demean the fallen, rather, find a way to have the victory appear as a point of instruction rather than as a failure for the fallen. These are the lessons which I found that OCS offers. These lessons have been my flotation in many a life's floods. The OCS that released me upon the world has been my greatest mentor. I simply cannot imagine what my life would have been like had I not attended OCS. It is the single most important milestone in my life's highway.

Virginia OCS had an extremely competent staff and they produced some outstanding graduates. Although great numbers of Virginia OCS graduates may not have become highly decorated combat veterans, a very large percentage took the principles of leadership with them to the civilian sector and excelled in every way. Some graduates did accomplish some extraordinary feats in a military career, for example, Jerry Headley of my own class. He joined the regular army and went to Vietnam after graduation and was awarded the Silver Star for valor on five separate occasions.

My bottom line is that OCS is one of the most productive programs in life. Whoever is able to be accepted should not drop out. Dropping out is the easy decision; maybe a sort of coward's decision; but I suspect that most people who make the drop out decision did so without thoroughly analyzing those forces that they think are driving them to the drop out decision. They find OCS different, difficult, unacceptable, or incompatible with their sensitivities so they rebel and quit. It is their loss.

With just a little bit more determination, resolve, commitment, they could have made it just like the hundreds before them. No one should quit. All candidates should hang with determination until they are thrown out for clear lack of ability to meet the high standards. Then, and only then, can they truly live the rest of their lives without a level of regret and wonder of what if I had not quit; how far could I have gone? I have heard this many times from those who quit OCS; and I have heard it many times from officers who eventually gave up their commission when life got complicated and they look back with some regret at having made that decision. But for me, quitting is not in my DNA.

One of my greatest pleasures was to go back and teach tactics to those classes that came along after mine. In all, over the years between my commissioning and my promotion to Colonel, I taught tactics to 19 different OCS Classes. My techniques were unique in that I created the scenarios from a blank sheet of paper. Of course, I had to achieve the learning objectives as specified by the Army but that was the only restriction placed on me. My trips to Vietnam and interaction with Navy SEAL Teams and with USMC Force Reconnaissance Teams gave me lots of ideas for scenarios which I employed freely. Later, Ranger School added to my bag of tricks and the difficulty of what I prescribed for the Candidates. A unique feature that they all appreciated was my personal participation in it all and that I was always on site, always observing and teaching, always making their life in the box a challenge and not a freebie. The only thing short of realism and toughness was the zip of live bullets. Others maintain that the courses I taught were legendary and are still discussed. When I was promoted to Colonel and given command of a Brigade, the Adjutant General suggested that it was time to step out of active participation in OCS and let some other young officer fill my boots.

COL Manley had obtained his commission at the FT. Benning OCS. Many years later he was honored by being inducted into the Benning OCS Hall Of Fame. He realized that his success as an Army Officer was the result of the actions of scores of fellow officers and Senior NCOs with whom he had interacted over his career. He composed a basic letter of thank you then personalized it for specific people and shared these with a number of individuals, known only to him. (In November, 2006, Douglas Manley, who had been my Senior TAC Officer in OCS, passed away and was buried on 22 Nov.) I was privileged to be one of those to receive the thank you letter from COL Manley. In addition to the personalization within the typed text, he also made a hand-written note in the margin as follows:

P.S. We go back to Class 6, and I have watched with pleasure your success after graduation. Your success has been based on a unique ability to be the "Ultimate Trainer." Your professionalism is always present. Once TAG said to me, "Childers is kind of wild, I am not sure he should get a higher command." I said, "Sir, if your sons had to go into combat with a NG Leader, who would you choose?" TAG answered, "That's not fair."

Platoon Level Crew Drill
SPC Allan Hughes

I am thinking it was Annual Training 1975 because we were at Camp Pickett VA. We usually did even years at Camp Drum NY and odd years at Pickett. Upon reflection, it was Pickett for sure because it was always hot and dusty and huge swarms of hungry mosquitoes to contend with. But I cannot complain because no matter what nature did to us, we always had more fun than anybody deserver.

It was an Armored Cavalry Troop that I was a part of Specifically C Troop, 1-223rd Armored Cavalry Squadron, 28th Infantry Division. Even though we were Virginia Army National Guardsmen, our higher headquarters was a PA AR NG unit. Known as the keystone Division because of the shape of the brilliant red shoulder patch, it was also known as the Bloody Bucket after the rather bloody battle it put up during WWII in the Hurtgen Forest.

Armored Cavalry is a very specialized unit in part because of the combined arms aspect of the Table of Organization and Equipment (TOE). Each platoon had 3 fully tracked tanks armed with a 90mm main gun plus a .50 Caliber heavy machine gun and an air cooled .30 caliber Browning machine gun that fired the 30-06 round of ammunition. Each Platoon had an M-113 Armored Personnel Carrier that transported a full infantry squad. Each platoon also had an M-109 carrier for a rotating base plate to mount a 4.2 inch mortar tube. The mortar could be fired mounted to the floor or it could be dismounted on the ground outside the carrier. Finally at Platoon level, there was a section of "Scouts", mounted in some version of the US Army ¼ ton Jeep. Each was equipped with communications equipment (as was every fighting vehicle in the Troop) and a pedestal mounted .30 Caliber Browning machine gun.

Our mantra was "Move, Shoot, and Communicate." Our primary job was Reconnaissance, Surveillance, Intelligence collection, and target acquisition. The Platoon was in fact a very mobile, lethal, mini-army and when the three Platoons were properly missioned, combined, they represented a formidable capability.

Now think about this; but not long. This simple little Platoon; this devastating, and sometimes unruly, group of death and destruction was commanded by none other than a Second Lieutenant.

Thankfully all the LTs were under the watchful eye of the Troop commander who was a Captain with typically 5 or more years of experience and had already learned the hard lessons and pretty much knew what would work and what would not.

Our Troop commander was a guy named Childers. He had been commissioned in Armor in 1964, distinguished Honor Grad of the VA State Officer Candidate School and prior to that had been through the enlisted ranks to the grade of Sgt, so by the time of this tale that I am going to get around to soon, he had about he had some 18 years' experience. So there was not much he did not know and in addition he was a very imaginative engineer within DOD. He was also just plain imaginative, especially in dreaming up training exercises, which is the crux of the tale which now begins.

It was called a tank section action drill. Here are the rules; hope I can get them right.

1. All tanks were parked in a large rectangular formation. Each Platoon's tanks were in a line. One tank section was up front then there was a space behind the first row then the second tank section, and a space behind the second row then the third section in a row. Adequate room was provided so that each tank turret could do a 360 degree sweep without banging into an adjacent tank.

2. All tank main guns were centered horizontal over the opening to the driver compartment

3. This formation of tanks were formed in a quadrant of a cross roads i.e. tank trails.

4. All tank engines were off as was all power switches of any kind within the tanks.

5. Each tank crew, 4 soldiers, were in a formation opposite from the rows of tanks parked across a tank trail. On signal, the front row of 4 crewmen had to sprint to their tank which was in the last row of tanks across the road. The second row of four crewmen had to sprint to the next to the last row of tanks across the road. And so on until 36 crewmen had sprinted to their tank, all having to sprint the same distance. And let me tell you, tankers are not accustomed to sprinting. They ride. And I am told that they always park in the shade.

6. Once all tankers had mounted their tank, started the engine, donned their crew equipment, established internal communications as determined by the Tank Commander, he would order the gunner to do a 360 sweep with the main gun and then give a hand signal that his section was prepared to begin the obstacle course.

7. The layout of the starting pattern took into account that if the front row was not the first section prepared to move, there was adequate room for a more rearward section to begin the next sequence. Ya know that a tank can do a neutral steer which means it can spin on a dime to whatever vector it needs to move out on.

8. The front tank section was to make a left turn, drive until they discovered their set of 6 each 55 gallon empty drums and conduct a sine wave path down the length, do a 180 and negotiate the six again and then return to the start point and dismount in formation at the front of the tank.

9. The second row of tanks was to make a right turn and the third row of tanks was to go straight across.

It was a hotly contested event to say the least. But of note was a bit of simple ingenuity by a single crew member. Everybody who knows me knows that I am a gambler at heart. Cards, golf, bowling, gunnery, driving. It is all a gamble and sometimes a small gamble pays big dividends. But here we are in the sprinter formation, looking across the tank trail at those tanks that are now about 125 degrees to the touch sitting in the sun and just BS'ing about what might save a few seconds. The whole thing was about the clock and about how well we could all do our individual parts besides just riding along. My tank commander was lecturing me on missing the barrels and how we are going to get into a march column to leave the start formation and I am half listening because I am looking at that main gun tube sitting over my driver hatch. Can I get directly into that hole without injuring myself to the extent that I cannot drive? I cannot mess this up. I think I can do it but I am not telling the TC 'cause he may nix it. I am thinking about the clock.

There are usually two entry hatches on the top of the turret. One is called the Commander's hatch and the other is called the loader's hatch. Typically the gunner will enter the Commander's hatch and then wiggle himself down into the gunner seat; then the Commander will enter and occupy the Commander hatch. Concurrently, the driver typically (depending on where the gun tube is parked) will enter the loader hatch and work his way forward and into the driver seat. But if I can squeeze my bod under that gun tube and into the seat, I can have my head gear (tanker's helmet) on and the engine cranked before the TC even enters his hatch. We are talking who knows how many seconds. And oh by the way, a little psychology; When the rest see my engine flaming and pouring out black smoke and their driver is not even in position, they are gonna panic.

So I told the loader what I was gonna do. "I am going directly into the driver hatch under the tube. Do not wait for me. Just take your position and tell the TC that I am in place. Of course he is going to feel the engine start up."

Guess Who Won.

So now you know the rest of the story.

The Porcupine Incursion. Fort Drum, NY
First SGT L.P. Hening

I think it was 1974. I was First Sargent and C-Troop was bivouacked for our two-week annual training at Fort Drum, NY. (This was our third trip to Drum for AT, alternating between Drum - even years - and Fort Pickett.)

One night Steve Graves had CQ (charge of quarters) duty. In the wee hours, I heard a noise outside my tent and got up to investigate.

I heard Steve's voice saying..."come on, get along, move it, move it, move it." Then I saw Steve with a long stick directing a porcupine through the troop area, herding him toward the officer's tent.

Captain Childers also heard the noise and came out of the officer's tent and saw what was going on. Childers threw open the tent flap and stood to the side as Steve herded the porcupine toward the tent opening.

By this time, most of the officers heard the commotion - the porcupine almost inside now and you've never seen so many LT's hauling it out of that tent in their skivvies!

Editors note...Being in the field at Camp Drum was better than staying in the barracks. To say the buildings were in "rough shape" is being generous. Little if anything had been done to maintain the barracks which were probably of pre-WWII vintage. Drum is now home to the 10th Mountain Division. Raccoons were also frequent night visitors to our site, making nightly checks on the trash cans in the mess areas.

Possum is not meat at all; I have tried it.
Carroll Childers, MG, ret

Maybe I did not cook it right. Surely I did though because I was once a "First Cook." The Cav Troop was at "Camp Drum", before it had the upgrade to Fort and there being not much else to do, I and the platoon leaders and a few hardy EM and NCOs decided to go for a middle weekend camping trip..........as if we needed more camping. We loaded a few 1/4 tons with a few C-rats and some items out of the mess hall, and of course lots of beer and ice; actually, mostly beer and ice.

Sometime during the night on Saturday , having been fishing and stabbing frogs all day, we were sitting around a rather large camp fire drinking beer and lying and suddenly someone cautioned "quiet, quiet; what's that noise out there?" So we all paused in mid drink to listen. Something was coming through the woods making a grunting sound. There are bears in NY State and we did have food out to be detected. And of course everybody knows that bears like beer.

So immediately flashlights came out and an axe, and there in the glow of a flashlight was the most unfortunate possum on the planet. With the first aggressive move by Sgt Salotti (yes, the guy who would later marry Janet and give her his last name) the possum played dead and Salotti cut his head off with the axe. Then began the discussion of what to do with the beast and eventually, maybe a beer later, it was determined that we owed it to the poor creature to consume him.

So it was field dressed, flushed with water from the water trailer, placed on a spit, and cooked slowly over the coals for about two hours. He was of course basted liberally with beer from time to time. Finally, it came time to eat it so bits were sliced off and passed around. Out of respect for my rank, they all waited until I began to actually take a bite. I mis-represented the delicacy by some factor that I am ashamed of. Nobody asked for seconds. The skeleton was found in the fire the next morning.

I cannot recommend possum spit roasted, although, it may be tolerable if baked slowly and covered in good Hot Pete Texas Chili.

Push Ups
SSGT Roddy Davoud

Never forget. I'm guessing spring 1981. Young African American kid comes back from basic and has maxed the PT test. Major or LTC Childers paid us a visit at a weekend drill and congratulated this young man on his physical prowess.

Oh and by the way, challenged him to a couple of pushups. Chest to wood, side by side for 30 or 40 reps until the young guy fell 1/2 rep behind and finally a full rep behind.

I didn't have a stop watch, but I suspect we were in the 60 in 60 seconds range! 41 vs 21.

Quite a demonstration.

Dress formation, AT Ft. Pickett 1968. 1st Platoon.

Dress formation, AT Ft. Pickett 1968. 2nd Platoon.

Dress formation, AT Ft. Pickett 1968. 3rd Platoon.

L-R: Captain James Bailey, 1LT Joe Lucas, 2LT Kenny Dias, 2LT Carroll D. Childers, 2LT Lommis D. (Dennis) Pounds.

Mounted pass in review, Ft. Pickett airfield, 1968

M-48A5 in its natural environment.

M203 range.

Track troubles.

Teamwork.

M88 set up for an engine swap on a M113.

Talbott at the wash point.

Childers (L) and SP5 Mike Thompson at the wash point.

The Mighty 88 in repose.

Drivers station on the M88.

Return wheel and track trouble.

Deploying out with the M113 for the super snow storm.

Common feature of Ft. Pickett...UXO 105mm howitzer round.

Cavalry formation advancing across open ground.

Camouflage applied.

Field camp, Ft. Pickett.

C Troop 223rd signage.

WW2 era barracks Ft. Pickett.

And then there's the paperwork... SGT Ed Heavener (Later COL Heavener)

Happy birthday 183rd.

Carrol Childers (MG USA Retired) graduated Class No. 6, VAANG OCS in August 1964. After serving 32 years in the Virginia National Guard, and holding commands at every level of the force, he was selected to command the 29th Infantry Division.

Instructing at a VAANG OCS cycle sometime in the 1980s, COL Childers instructs candidates on operational planning and execution.

Riding the Huey Skid
Carroll Childers, MG, ret

I forget which training year it was but I do remember we were at Ft Pickett VA. We were C Troop, 223rd Armored Cavalry Squadron. LTC Cliff Boyer was the Squadron Commander. The CAV and the Aviation had been having some good natured adversarial relationships as usual and it was hard to tell which group was ahead; but it was close. One day, I think it was Saturday afternoon and the training tempo was down somewhat while maintenance was being done in preparation for the second week of high tempo training. I got a call from the Aviation Commander, one James Holden, actually a super soldier other than that he was the enemy. He had been awarded the Silver Star for valor in Vietnam as a gunship pilot. I never did know the precise story for his award but it was generally recognized that he was fearless and a very, very good pilot. His personality was self-malleable, that is, he could turn from a smiling laughing friendly guy to a stone cold mother-flogger in a millisecond. Maybe it was because he had earned the Ranger TAB?

The phone call was certainly unexpected and the purpose was nearly unbelievable. He was inviting me and my XO, LT Buddy Deverell, to accompany he and his XO on an aerial recon of the post in support of training the next week. He said they would be locating small LZs and checking them out for safe landings. Back then, early 70's, these pilots returning from Vietnam could shoot LZ's that would simply not be allowed in another 5 years. They were good and they could get in and out of the most impossible looking places, barely clearing main and tail rotors many times. Dragging the skids through tree tops was not an uncommon occurrence. So he offers us this free tour, saying that knowing where these LZs are might be tactically important to us as we plan our training.

Sounded logical so I accepted, frankly, with some considerable excitement. Here was perhaps the best pilot in North America and he was going to give me some exciting inserts and exfills.

Buddy and I reported to the LZ at the appointed time rearing to go. So off we went. Jim Holden did not show, rather he sent CWO Macintosh as the command pilot and a unique pilot named Bob Turner as the left seater; no crew chief. All of this gave me some suspicion but still I went for it. Buddy and I took jump seats behind the pilot/co-pilot but they were arranged as far back as possible; not the most common arrangement I was accustomed to. We did indeed shoot several tight LZs. After 5-6 of these, they landed and there was a lot of small discussion between the pilot/co-pilot.

Buddy and I had earphones on and could hear the conversation and it seemed to be about engine power. I am on the side Port side and Buddy on the starboard side. After sitting there for several minutes, Bob dismounted with an extension of comm cord and walked around the front of the chopper to the port side where I was sitting and motioned me to get out. I am falling for the trap. I get out and he tells me to follow him and to motion for Buddy to get out. So the three of us walk forward just outside the rotor wash "so that we can hear better." Now that gave me a clue that all is not on the up and up. He said that the engine is exhibiting some strange behavior and that they need to test the lift with minimum load and that he and we three passengers will wait on the ground until Mac can take it through this supposed lift test.

Then he gives Mac a signal and Mac lifted off the ground just a little then plunked back down and pretended to fool with his dash some. Then he told Bob via the extension cable to come to him. Bob turned to me to tell me that Mac needs his assistance and turned to go back to the Huey. I went with him, uninvited.

This was not their plan. So when Bob reached the Huey, Mac and Bob had a mini conversation via comms, pointing and gesturing. Then Bob motioned to me and I followed him back to where Buddy had waited outside the rotor circle. They went through this act a second time. Again I followed. Act three. This time instead of Bob going to the pilot window, for the gesturing routine, he went to the starboard side of the Huey and stood some several paces from the Huey. I went to the Port side and stood about as far away as Bob was standing. Buddy had stayed just outside the rotor circle again.

All of a sudden Bob sprints for the Huey and Mac pops it off the ground. Bob literally dives into the open door landing on his belly on the floor. The Huey continues to climb as I sprint toward it. I make a tremendous leap, contacting the skid with the left hand. The chopper is climbing. I slap a grip on the skid with the right hand and swing my right leg over the skid and lock its ankle with my left knee. I am on that craft. It scrapes me through the upper branches of the trees as it clears. Buddy is standing in the LZ, probably wondering if I am in on this trick or not!!

The Huey is banking right, I am hanging on for dear life, and bob is holding on to the load rings in the floor to keep from sliding out. All of this is extremely against all regulations. When the Huey levels off, Bob crawls over on his hands and knees thinking he will be waving goodbye to two idiots in the LZ who have no idea where they are but where ever they are, they have a long walk home. The look on his face when he saw me hanging onto his skid was incredulous; worth a million dollars. He screamed "Hey you SOB, get off of my skid!!" Then he laughed, "You crazy SOB, hang on while I get you inside. So he put on his safety belts and got me pulled in.

They made a couple of low turns around the LZ where Buddy had been stranded, waved good bye, swapped middle fingers with him, and flew off low making several turns as they went to confuse me as to where Buddy might be stranded. But I knew exactly where he was as I was familiar with every tank trail on the post and I picked up a critical junction shortly after leaving him. They flew back to the Airfield feeling pretty proud of the trick they had pulled even though they only left the XO.

Bob put on his stone cold face and said, "We will deny this event and I have more credibility than you do." Bob had one tour in Viet Nam as a grunt and a second tour as a helicopter gunship pilot. I responded "That's ok Boy (we called each other Boy); I don't hold grudges, I just get even. A piece of your hind quarters will adorn my command track before the week is out."

Jim who chose not to personally participate in this caper, but I am sure he dreamed it up, and I were both Captains in the 70's. I was always a so called M-Day soldier, or Title 32. Jim became a full timer or Title 10. He eventually rose to the position of State Aviation Officer, then to the Chief Of Staff to the Adjutant General. As we both became LTCs, I went off to the resident course of Army War College which put me in a good competitive posture to continue to rise through Colonel and into Flag rank. Jim eventually was selected to serve as the 116th Brigade Commander in I think 95. I went on to make Brigadier General in 1993 and was the Assistant Division Commander for Maneuver and became Division Commander 96-99.

The 29th did some great things in training and evaluation and gained generous recognition for the Division to the extent that the Chief Of Staff of the Army, General Dennis Reimer at the time, spent an entire day in the field with us at Annual Training viewing our unique and challenging training program which we called Operation Chindit.

Supply Officer Turn In
LT Bryce Bugg

AT 1972, (Camp Drum) I was the supply officer and accordingly, I had to turn in all of our gear before we could clear post.

I walked into the building where this was supposed to take place. Holy Toledo (I cleaned that up)! The line was reallll long.

I glanced up toward the lady who was in charge (and doing the checking of everything) and to my delight and surprise, it was Mrs. Holloman. Mrs. Holloman's daughter married one of my groomsman. Mrs. Holloman loved me and I her.

I walked around the table and came up behind her and kissed her on the cheek. THAT, she was not expecting! She jumped about a foot.

When she saw me and realized why I was there, she grabbed the papers outta my hand signed them, returned the kiss on the cheek and told me to get going. I did!

I had several offers to join other units as I left with a big smile.

I ain't what you know, it's who you know. So true.

Tank Incident, Nottoway River Bridge, Camp Pickett
Carroll Childers, MG, ret

The Cavalry conducted training, typically at Ft. Pickett, but sometimes at the Dove Street Armory, one weekend each month, every month of the year including December if the weather was permissible. Typically December and January, being inclement weather months, the weekend drill would be conducted in the Dove Street Armory focusing on administrative requirements and Civil Defense training. The other 10 weekend drills would have been referred to as a Multiple Unit Training Assembly-5 (MUTA-5), meaning Friday night, Saturday, and Sunday. Each 4 hour period available in these 3 days were a UTA, thus 5 UTAs, thus MUTA-5. Details of this are the subject of another story.

Understand that there are more than 5 x 4 hours of available training in this long weekend but because C Troop was a heavy vehicle centric unit, maneuver training after nightfall was considered to be a risky event and therefore the kind of training conducted was selective and conducted with great care. Normally the night field training was limited to night range firing or night road march because a heavy measure of safety could be applied to these limited events with success.

This particular story concerns a night road march in which a near disaster was turned into an exciting and unique memory. C Troop was a unit in which few soldiers ever walked from point A to point B. Sometimes we used roads (gravel) and trails (dirt/mud) and sometimes the travel was cross country through woodlands and fields, making our own paths using the power of the tank to blaze trails sometimes. In the troop there were 9 combat tanks (M-48 or M-60), 3 Armored Personnel Carriers (APC) M-59 or M-1113, 3 Armored Mortar Carriers, M106, 1 M-577 Command track, one M-88 Tank retriever, X number of tactical vehicle wreckers, 4 medic vehicles, 15 Scout vehicles (M38A1, M151, or M114; depending on the year concerned), and an unknown number of 1/tons used as Platoon Leader command vehicles, HHC administrative vehicles, and ones we had made a midnight requisition for special uses as necessary. Altogether the fleet was some 50 wheeled and tracked vehicles. Most of these mounted tactical radios for instant communications for command and control.

A night road march is not just a ride in the dark, although, it is certainly that. What else is it? Well it is a carefully planned military operation in which the entire fleet navigates a prescribed path over a long duration (4 hours minimum), maintaining contact between vehicles of a given march unit, reporting issues as they arise, reporting specific identified check points along the march route, making prescribed halts and excursions built into the plan, and maintaining safe separation between vehicles.

On the night of this story, it was particularly dark. Starlight was the only illumination and because the planned path was on tank trails, both improved and unimproved at times, the only vehicles that were not visually impaired by the dust were lead few in the column. Vehicle commanders and drivers were equally responsible to observe and safely follow the vehicle in front of them via the only instruments and methodology available at that time which was a tiny pair of very dim illuminators which were referred to as "cat eyes." If the pair of cat eyes on the vehicle in front of you appeared as a single eye, your gap between your vehicle and the vehicle in front of you was too large and you must carefully close that gap until two eyes appeared again.

So the challenge was to maintain a consistent speed which would minimize speeding up and backing off because as you did this in your vehicle, imagine what happens with the vehicle behind you, and the one behind that vehicle and so on down the line for a column of 50. This is a terrific strain on the driver and the vehicle commander over time. Moonlight makes a tremendous difference and of course, today, with night vision devices, this is a relatively easy mission. But this is why we did this drill, to become proficient and confident.

On one night many years ago at Camp Pickett, C Troop was conducting a night road march. As the C Troop commander, my driver (Mike Thomas) and I were at the head of the column in our trusty M151 and had been under way for a couple of hours and looking forward to a few PBRs at the end, when the radio crackled a message without the formality of first making contact. "Six (my call sign), the column has halted. You better get back here right away."
Well, the column behind the caller had indeed halted but everything between him and me was still moving. I think it was probably the scariest radio call I have ever received because of what was not said. Our Operational Order (OPORD) had a provision to halt the column and I executed that code. I can't recall if I ordered headlights on or if Mike just did it on his own initiative but I know he automatically did a combat 180 turn without a backup and went roaring to the rear.

As we shot to the rear, the column of vehicles had alternatively taken left shoulder/right shoulder positions on the gravel road. On the way back I was getting more info on the issue so by the time I reached the incident area, I knew that Sgt Talbot's tank had ripped down about several tank lengths of guard rail on the bridge across the river and was hanging precipitously on the 6x6 edging of the bridge planking. I called the Motor Sergeant, Maurice Harver, but he had monitored the call to me and was already on his way from the rear of the column.

It was quite a scene. The command jeep headlights shown on the front of the stricken tank and shortly Harver arrived from the rear and his lights lit up the rear. The crew was still in the process of getting out carefully and they had cut the engine. The tank was essentially balanced with its right track at midpoint on the 6x6 timber which served as a curb for the wooden planking and as a structural member to fasten the guard rail to. I had the distinct impression that the tank was lightly teetering back and forth as though the weight of a crewman moving around was disturbing this balance.

As it turned out later, the tank was near exactly in the middle of the river. My estimate is that if it had traveled another foot or two, it would have toppled off the bridge and with the inertia of a rolling tank at the altitude the bridge was above the river, the tank would likely have had the geometric orientation to have turned 180 degrees and went in top first and sunk turret down to the bottom of the river. All escape hatches in the turret would have been blocked. The driver belly escape hatch would be useless as it would have to be lifted in the dark under water against the pressure created by the depth of the river. The tank had been stopped by sheer providence or the Grace of God.

I don't think the crew even knew they were on the bridge. They only knew they could no longer see and they stopped.

Harver and I talked briefly. "I will drive it off Captain. I am the best tank driver here so I am going to do it," Harver stated. I think it was his way of giving the Captain an order.

"I am going to drop the driver escape hatch before I crank it up. If it should go in the river, I will get a replacement," he added.

"Don't rev the engine when you crank it as the inertia of that could create an unbalance, it looks like it is about to fall over now," I cautioned Harver.

So that was the recovery plan. I have thought about it many times since and wondered why I did not call the M-88 forward and use a long cable and tow the tank back to a more stable position on the bridge. I guess I had a lot of trust in Harver's confidence that he could successfully remove the tank from its precarious position. This was a case of agreeing to a course of action. Who knows, the longer the tank sat where it was, loading a bridge in a unplanned manner, the more likely that something unpredictable could evolve; like something yielding somewhere which would cause the weight of the tank to shift slightly and we had no idea in the dark what we were dealing with. It seemed like the quicker the better. This was a bad situation that could not get better with time.

So, Sergeant Harver mounted the tank very carefully, worked his way around the left side of the turret on the sponson boxes (storage containers on the fenders) and into the driver hatch under the gun tube. I heard the escape hatch making a steel on steel screech as Harver unlatched it, followed by the splash at it impacted the surface of the river to sink forever. The engine roared to life followed by the mechanical "clunk" and the sound of the tank being shifted into reverse; torque being transmitted from engine to transmission and the movement of final drive and the tank eased back to the centerline of the bridge. I took a really deep breath.

I got on the radio and announced, "Situation resolved, there is no longer a guard rail on the right side of the first half of the river bridge. We will resume the motor march as soon as I return to the head of the column. Stand by."

It was only while this story was being discussed during the writing of it that I learned "the rest of the story." There had been "passengers" in this particular tank this night, beyond the normal 4 crew members. There were two additional riders this night; two admin soldiers who wanted the thrill of a night road march in a tank. They had no idea how close they came to the first and last ride they would ever hitch.

Tank Rescue
Carroll Childers, MG, ret

C Troop was conducting Maneuver exercises at Camp Pickett, during Annual Training as I recall it. Maneuver exercises are a free play event wherein one element attempts to gain the advantage over the opposition by conducting an unexpected move from point A to point B, or perhaps the movement is over terrain which the opposition has judged to be impossible/improbable and it is discounted, thus not monitored by Scouts or other resources.

This particular exercise had hardly gotten underway before it was interrupted by a tremendous micro-storm. Both wind and rain were tremendous as was thunder and lightning. It takes judgement to make the call but it is not unusual to temporarily suspend training until such bad weather is over, especially when there is lightning. In this instance there was a break of about 30 minutes in training and we then resumed. The maneuver would be through a moderately wooded area, interspersed with several creeks. They would be passable if done cautiously and would totally surprise the opposing force. One tank was maneuvering near the base of a ridge which rose from a creek on the left up to the crest of the ridge on the right.

Neither of the tank crewmen paid much attention to the fact that a large popular tree had fallen across their path in the previous storm. It was very innocent looking. It was the product of a logging operation several decades prior, in which twin trees had grown from the stump of a harvested tree, and matured. This stump root mass happened to be located at the head of a deep creek and when the wind blew this twin trunk over, the root ball was ripped out of the ground leaving a great hole at the headwaters of this uniquely located creek. The runoff from the previous downpour was funneling into that root ball hole and washing on down the creek. The surrounding terrain, hills, poured runoff into this little creek but again, no one paid much attention, after all, it was over several yards to the left.

Then fate made its move.

The tank driver drove cautiously up on top of the downed twin trunk popular tree. Trying to be as silent as he could, the driver was not crossing the obstacle aggressively so as the tank briefly balanced itself on the twin trunks, the weight of the tank, 96,000 pounds, broke through the popular bark and discovered a very slick surface of green, juicy wood. Remember, this tree fell up hill and to the left is the creek. Without warning the tank began to slide down it acting like twin greased rails, peeling bark as it went and gaining speed. It was a runaway train not to be stopped.

Down it went, moving perhaps 80 feet in only a few seconds, bulldozed its way through the root ball and dropped squarely into the creek in a nearly perfect fit. Water was over the engine compartment. A tremendous geyser of water was thrown up by the tank plopping into the creek and the driver was suddenly under water. The crew compartment began to fill rapidly. The gun was forward so the driver had no real challenge to climb out of his hatch and up on top of the turret. The TC and Loader were standing in their hatches and they had no real danger but the TC did have to evac his hatch so that the gunner could exit as he sits below near the bottom of the crew compartment.

Within a few minutes the whole platoon had heard some discussion on the radio net and had located the stricken tank and had assembled nearby and were discussing the situation. I arrived perhaps 10 minutes after the main event and was in overload by people giving me their version of what had happened. The lesson was clear; never attempt to drive slow across a newly downed double trunked tree on a hillside. My mind is working on how to recover that tank. I am not sure that the M-88 is up to this task. Someone had said that the Motor Sergeant had been given a SITREP and the 88 was en route. Radio communications with highly effective military technology was one of the great benefits of the Cavalry.

The layout of the stricken tank in the creek was like a cork in a bottle. There was a few feet space between the rear of the tank engine compartment and the earthen walls of the dead ended creek. Water from runoff was still seeping into this pool of muddy water and then oozing through the tank track system to continue to flow downstream. I suppose that if we had waited long enough, the pool of water may have drained away and we could have assessed the situation and had more ready access to the recovery points on the rear. The engine was off and the crew compartment was still full of water. Who knows what might have been wrong. What to do? No doubt that we collectively had an intense desire to remove this cork from its bottle so that whatever damage may have been done to the tank, including water damage, could be corrected ASAP.

The 88 arrived. The Motor Sergeant walked around the site a time or two looking at everything. "There is too much clutter in front of the tank so we cannot move it that direction at all so we are going to have to yank it out from the rear," summarized Harver. "Let me get the 88 positioned at the rear and see how it looks and how stable that ground there is, we don't want to end up with the 88 sitting on the back deck of the tank."

Harver ground-guided the 88 into a very precise position at the rear, several feet above the tank's rear deck. He was kneeled down looking at everything, smoking a cigarette. I was kneeled down with him. We probably should have been praying because this tank was embedded in the earth and the only way out was up and back and the source of power was the 88.

"Captain," Harver offered in his gravelly voice, "I believe it will come out of there, if we can get the tow bar attached to the aft of the tank. Because all of the tow points are under the muck, we might be better off to hook two to the 88 and only one to the tank. Hook it to the tow hitch." I don't know who he was talking about as he was sort of thinking out loud. He stood up and began to unbutton his uniform blouse. I reached over and held the button he was working on;

"You hook the bar to the 88 and I will hook it to the tank."

"Captain, you don't have to do this," he said, "but I know if you have made up your mind, I know I can't stop you."

"Well Sergeant Harver, one thing I know for sure, if anybody gets taken out of action in this retrieval, it cannot be you because you know more about this kind of thing than anyone else."

With that I unfastened my web belt with all my field gear attached and placed it on the ground with my headgear and emptied other things out of my pockets, assigning security of it to my 6D. Then I turned back to my field gear where I always carried a number of sets of ear plugs and retrieved two sets; one for my ears and one for my nostrils. I did not relish that muck entering those critical orifices. I did not relish entering the pool of muck and I had no idea at all if I could; manhandle that tow bar, find the tow hitch, and avoid getting myself entangled in something. Nor did I know if the geometry of all the piece parts would allow me to insert the eye of the tow bar into the tow hitch and then to close it, all in water which that was so mucky that no light penetrated. Didn't plan on opening my eyes anyway.

The clock is ticking and the situation is not improving with time. Harver shouted over the noise of the 88 that the tow bar was attached to the 88 and that "my best guess is that the eye will be very close to the tow hitch. Be careful Captain."

I slid down the bank into the pool of muck, looked up briefly at the 88 which seemed to me might just slide over into the hole with me. I positioned myself on the left of the tow bar, took a deep breath, and submerged to feel things out. It was years later that I wondered why I did not have a rope tied around by chest and under my armpits, just in case, but at the time it never occurred to me. My hands followed the towbar down to the eye at the end and using my left hand to hold the eye as a point of reference, I searched with my right hand for the tow hitch. Unbelievably, there it was just as the ole Motor Sergeant surmised, it was so close to being connected that I could not believe it. This is going to be deceptively easy.

The eye moved on a perfect arc and clunked into the hitch, without taking my fingers with it, and I moved the pivoting latch down to entrap the eye. I recall thinking, "It's hooked." I don't think I had been submerged more than 20 seconds when I stepped up on the hitch and broke surface. Fifty or more questioning faces staring down at me like a family of prairie dogs.

"Did you get it hooked Captain?" shouted Harver over the 88 engine noise.

"Got it Sergeant Harver," I responded. "Let me get out of this muck hole and you can yank it out." The ring of somber faces turned to a ring of smiles and cheers of some relief.

I cleared the immediate area as Harver expanded the diameter of the onlookers and when it was done to his satisfaction, he climbed on to the 88 and exchanged some unknown instruction to the driver, which I think was John Mull. He gave a thumbs up and the 88 roared to life as it lurched rearward firmly attached to the stricken M-48, dragging it like a bull dog would drag a stuffed toy. They drug this mudball monster about 4 lengths of the tank to be sure that it would not have any potential to re-enter the trap. Harver executed the hand and arm signal for land vehicles to stop, clasping his hands in front of his chest with a slight movement of urgency. Three smiling faces on the 88, ecstasy among the onlookers. C Troop had triumphed again.

Personally, I was indescribably relieved that no one was injured, and that the retrieval was effected on our own. C Troop did not have to go hat in hand for Post assistance by MATES. I was also forever impressed with the capability of the M-88. I can never forget the up close and personal demonstration of sheer power and the ease which the tank was yanked from its muddy mire.

Tank Section Training Accident at Pickett
Carroll Childers, MG, ret

It was a weekend training MUTA-5 at Ft Pickett, as I recall in the spring or summer. We had convoyed from the Dove Street Armory on a Friday night, spent the night in field bivouac, and arose at dawn on Saturday morning to conduct tactical training. Training was going well as the kinks were worked out. The National Guard conducted weekend training basically once a month so it always took a few hours on a Saturday morning to once again get into the swing of converting from a eight to five citizen in an office or factory or other environment to the environment of the soldier in the field. To multiply the environmental conversion by a large factor, we were maneuvering on unimproved, sometimes unexplored, terrain in vehicles designed for war. Only the reconnaissance elements were in vehicles remotely similar to civilian vehicles, other than being equipped with pintle mounted machine guns and being four wheeled drive and capable of sometimes unbelievable mobility cross country. Essentially everybody else in the Troop traveled in full tracked, armored vehicles such as M113 Armored Personnel Carriers, M-106 Mortar Carriers, and the King of Battle, the 50 ton M-48 Tank.

The accident occurred on an M-48 so I will focus on the tank. The tank configuration includes the tank hull, the lower portion of the armored vehicle to which there is a suspension and the track system which provides the means to give the tank mobility. The hull also has an engine compartment which hosts, in the case of the M-48, a gasoline engine of some 1200 HP. Fuel Tanks are also within the hull as is the transmission and the various mechanisms which connect the engine to the final drives which cause the tracks to move the tank. Then there is the Turret. This hosts the main gun, a 90mm high velocity cannon, as well as a .50cal M-2 Heavy machine gun and a .30 Caliber machine gun mounted in the turret coaxial to the main gun. Various fire control elements are also present, such as optics to aid in aiming the main gun and traversing mechanisms. Finally vision blocks provide sight for the various crewmen who need to observe while buttoned up inside the tank so that the enemy cannot enter it uninvited. The whole thing is mounted on the hull via a geared turret ring so that it can be rotated endlessly in either direction to allow the main gun to be fired 360 degrees.

The tank hosts four crewmen as follows: First the Driver, who has a tiny compartment in the forward part of the tank hull. He essentially sits below the main gun and has full and explicit control over the mobility of the tank. As he moves the steering control he causes braking of one or the other of the tracks to effect turns; brake left track to turn left or brake right track to turn right. He can also do what is called a neutral steer wherein one track will engage forward while the other engages rearward. Using neutral steer, the tank can be made to spin a full 360 degrees or more. The old tanker's rumor was that the transmission would only tolerate 50 full spins of 360 degrees before it would fail. I never believed that story and never saw that end result in 25 years of service as an Armored soldier.

The driver had a hatch which he could close and lock internally to protect himself from direct fire. He also had an emergency escape hatch in the floor which he could unlock and drop to the ground, thus giving him a very uncomfortable port to escape through. He also had the capability to communicate with the other members of the crew via a radio intercom system using a headset, microphone, and control box. If he was locked into his compartment he had vision blocks, much like periscopes, with which to observe outside, but only in the forward direction. He had very little field of view to the rear of the left and right fenders of the tank track. To assist him in maneuver when the tank was buttoned up, other members of the crew, using their vision blocks and the intercom, would assist the driver with direction.

The turret of the tank hosted three crew members; the tank commander (TC) the gunner, and the loader (for the main gun). The tank commander fought the tank and ensured that the other three crewmen were trained and could do their jobs proficiently. Most of the time the TC stood on his seat as such that essentially he was exposed to line of sight from the waist up. This gave him an excellent 360 degree capability as quickly as he could rotate his body or head. If he were required to be buttoned up, meaning down inside the turret out of line of sight, he would be sitting on his seat and viewing through a series of vision blocks so his capacity to view 360 degrees was complicated and time consuming as he would have to move from block to block. He was continuously in contact with the crew via intercom and in contact with other vehicles of the Platoon via the radio.

The Gunner was responsible for engaging targets with the main gun and in this regard he communicated with the loader as to what kind of ammo to load into the gun breach. There was a family of ammo (by type) to choose from, for example, High Explosive or Armor Piercing. The Gunner's position was aside the gun, to the right of the breach, and below and forward of the seated position of the TC. The Gunner of course had Radio/intercom.

The final crewmember was the loader who also functioned as the assistant TC and if the TC became inoperable, the loader would assume that duty and also attempt to continue to serve as the loader. The loader was located to the left of the main gun breach and he had two possible positions from which to operate depending on the situation at hand. If firing was occurring, he would be on the floor of the turret. The floor rotated with the turret so the loader was always right there convenient to the stowed ammo and could load as demanded. His alternate position was to stand on a ledge and protrude from his loader hatch very much like the TC protruded from the TC hatch.

There is also an externally mounted hand held telephone instrument which, via an extendible connector cord, allows non crewmen who are accompanying the tank across country to communicate with the internal crew. These outside eyes were useful for assisting the crew in the employment of the tank as a fighting system alongside infantry.

All hatches were a thick steel flat plate which had a spring assisted hinge on one side and a latch to lock the hatch closed from the inside on the other. Another feature was a spring loaded latch affixed to the tank turret whose purpose was to engage a catch on the hatch and hold it firmly open so that it would not swing free and be a danger to the hatch operator (TC or Loader). The hatch had a hard rubber gasket which would, when closed fit down into a u-shaped groove cast into the surface of the roof of the turret. Keep this groove in mind as it is the essence of the story as I get to it.

The 4-man crew of a tank is a team. If the tank gets stuck, all four assist in retrieving it. If it throws a track, it is not the job of the driver to reassemble it; but the whole crew. When it is time to turn the tank in, all 4 assist. When it is time to camouflage the tank for tactical concealment; the whole crew. And on this particular training event, all tanks had to be camouflaged which meant cutting lots of small trees and limbs and selectively wedging these into various features on the exterior of the tank so as to make the tank blend with the wood line.

All the vehicles had been successfully camouflaged, inspected for quality of camouflage, and approved for the conduct of follow-on training. The signal to mount up was given; the troops scrambled aboard, commo was established, and the order to move out given. The unit had not even gotten into march order along the planned route when an emergency call came on the Platoon frequency for the Medic.

Spec 4 Fred Brunner had been a tank loader for this movement and he was in deep trouble. I don't think I had ever seen fingers more seriously mangled in my life. It happened so quickly and innocently and I was particularly concerned when I saw his fingers, on both hands. Fred was a fresh new Mechanical Engineer who depended on such features as fingers to do such engineering tasks as drafting and operating a slide rule. I was deeply concerned.

What had happened? How did this occur? What had we done that we ought not ever do again?

Remember that rain groove in the exterior of the turret roof? Remember that latch that engages the loader hatch and holds it open? Remember all that foliage that we had stuffed into various features of the surface of the tank? Someone apparently stuffed a piece of foliage into the area around the loader hatch so as to unlatch the hold-open catch. As such, the hatch was no longer secured in the open position. Then, as is often the case, when leaving an assembly area to form up in march formation, the tank will surge and stop and surge and stop or run over a log or hit a ditch or any number of things to modulate inertia of the tank and contents.

Then at some point, Fred decided to exit his hatch and he reached up and got a grip on the hatch opening to pull himself out for exposure from the waist up. So his fingers spanned the u-shaped rain groove just at the right time for the hinged hatch, now unlatched and free floating, to be accelerated by the tank movements and cause the hatch to close on his fingers. Thankfully, his head was not emerging from the hatch. The fingers were pressed, backwards against the normal movement of them, into the rain groove.

As it turned out, the groove was a near perfect match to the distance between joints/knuckles of his fingers which minimized permanent damage to Fred. He recovered rapidly. Fred retired after many years as a Command Sergeant Major and, unfortunately, shortly after retiring passed away.

The Drive Home
LT John Terry

I don't remember the particular year, but Capt. Deverell was the TC and Lt Redfern was the XO. It had been a tough AT. Several wrecks, one involved Clarence Taylor saving a trooper's life by holding his head out of the water until help could come and overturn a jeep. I think it was the same AT that the cheese caught fire on the mortar range. A trooper was burned, thankfully not serious.

We were in convoy leaving Ft. Pickett and returning to home station. Duke Karnes was my driver. He was smoking a cigarette and I was scanning a Playboy looking for good jokes. Duke says the engine cut off. I told him to brake slowly and pull over to the side. The area alongside the road appeared flat. But when we pulled over we discovered that there was tall grass and a considerable slope. It felt like it was 45 degrees.

So there we are on this slope with Duke trying to slow the jeep down and right in front on us about 5 or 6 yards is this oak tree. Maybe 2 feet in diameter. I see my life flashing in front of me. I was still young and it was a short story. But before we could hit that tree the jeep flipped upside down! Just jumped up and flipped! (a common issue with 151s due to the independent suspension) The only thing that saved us was the antenna mount. My jeep had a 3 foot extension off the back of the jeep. That extension prohibited the jeep from crushing Duke and me. Now Duke and I are upside down and I am calling the C Troop commander requesting that he stop the convoy. He called back and wanted to know what was going on and I reported that there had been an accident. Us.

We were upside down under the jeep and gasoline starts to leak on us from the tank. You may remember that the tank is under the driver's seat. This caused some panic. Duke had been smoking before the flip and we didn't know where his cigarette landed. After getting out from under the jeep, we started to police up the various articles strewn about. We especially moved the remnants of some empty cans of adult beverages that were left over. The cooler, cans, Playboys anything that appeared unmilitary.

Capt. Deverell drove up and if looks could kill I would not be writing this report. But in all fairness, I'm sure that he was asking himself, what else could possibly happen. After all we were on our way back to home station.

Looking at the wreck we determined that since the jeep was an M 151 (which can be air dropped) with each wheel having independent suspension the two inside wheels collapsed. Lucky for us we had that extra-long antenna mount or we would have been crushed!

The commander called the MP's at Ft. Pickett but was informed that since we are off base it was not their area of responsibility. However they called the State Police and we were to wait for their arrival. Within 20 minutes a State Man showed up. Young officer, maybe 6 months on the job. He informed me that if it was less than $250 damage he did not have to write it up. I assured him that it was less than $250. He left us and we awaited a truck from Ft. Pickett to tow the jeep back to OMS.

To this day I don't know how much damage was done to that jeep. Shattered windshield and badly mangled antenna mount. But no one was hurt! I get to tell people that my driver flipped a jeep on me and I lived to talk about it. Of course I don't add that he was following my directions.

This is another example of how much extra our AST's had to work because we were the Calvary!

The Pleasure of Armored Cavalry
Carroll Childers, MG, ret

The Virginia National Guard thought enough of my performance while attending the Officer Candidate School (OCS) to recognize me as the Distinguished Honor Graduate. My commission was as a Second Lieutenant, Armored Cavalry. My first assignment was as a Platoon Leader in C Troop, 183rd Armored Cavalry Squadron, 29th Infantry Division.

There is no assignment in the Army more challenging nor more enjoyable than the Armored Cavalry. At the Platoon level, a mere Second Lieutenant is issued a combined arms team including Armor (tanks), Infantry (mounted in fully tracked Armored Personnel Carriers (APC's), indirect fire assets in the form of 4.2 Mortars mounted in a full tracked armored vehicle (similar to the APC but outfitted internally with a base plate for the Mortar), and a Scout Section (Reconnaissance and Surveillance) mounted in the Jeep of the day with a pedestal mounted .30 caliber Browning machine gun. Every vehicle had two way radio communications for command and control, and of course the soldiers who occasionally dismounted (infantry and Scouts) also had a back pack radio of the day.

Initially, our tanks were the 27 ton M-41 Walker Bulldog tank with a 76mm main gun; so I would have to presume that we were in fact a "Light" organization. The mission of the Light Armored Cavalry Squadron/Regiment would be to conduct early reconnaissance and security operations in support of contingency operations. We trained hard and seriously, even though the Regular Army had little intention of recognizing or providing much support to ensure that we could accomplish our mission. From commissioning up to and through command at the Troop level (O-3) I was in the Cavalry from 1964 to 1976, having commanded a Troop from 2 Jan 72 thru 31 Jan 76.

Cavalry Troopers are unique soldiers. They honestly believe that they are the best and there is little that they cannot do; but they continue to aggressively seek out that which they cannot do so that they can figure out how to do it. They learn some good lessons, some hard lessons, and they make both friends and foes by some of the things they do in their pursuit of excellence. One thing they love is competition in field exercises against other units, particularly if the other unit also has a lot of esprit. Sometimes the outcome is not popular especially by those who cannot take a joke; or maybe these unhappy objects are too prudish.

I recall an event in which my Troop shut down flight operations for an entire aviation unit (Company sized). We were playing tactical games that included the Aviation. Their LZs were tactical and were relocated at unpredictable intervals but we found out through surreptitious means that they planned to move back to the base airfield for one night as a "tactical diversion." Actually it was their way of getting back in for a shower.

Our objective would be to take control of the aircraft log books, thus grounding them. By regulation, they can't fly without the log book. They can do a lot of silly things but when it comes to violating aviation regulations that is where they generally drew the line.

The Cavalry Troop was made up of 3 line Platoons, each with a scout section mounted in jeeps with 30 caliber pedestal mounted machine guns. In addition, each platoon had a rifle squad normally mounted in M-113 personnel carriers. The plan was to cause a diversion by having the three Scout Sections drive through the hangar and billeting areas around the airfield blasting with machine guns (firing blanks) and tossing simulators and smoke grenades while the dismounted infantry made a raid on the line of helicopters, taking the log books from the cockpits where they were normally kept after crew maintenance.

The Scouts created quite a commotion, so bad that the aviators got totally pissed off and called the city of Blackstone and asked that the police respond to the "dangerous activity" by these people in gun jeeps. By the time the sirens and flashing lights showed up, the scouts and infantry had long gone………..with all of the aviation company logbooks.

About 0700 the next morning I had a visit from the Squadron Commander, his XO, his JAG officer, and the

Aviation Company Commander. I invited them to join us in breakfast.

In the sternest manner I had yet seen from the Squadron Commander, he replied "This is not a social call Captain Childers. Did you take the Aviation Company log books last night?"

I replied, "Sir, I can assure you that I did not take anybody's logbooks. What would I possibly want anybody's log books for?" (I, of course, meant me personally)

The Commander responded, "You know that they cannot fly without log books?"

"I had not really thought about it that way sir. Let me look into it. It could be that some of the men decided to borrow the log books to examine them for some evidence of the Aviation having illegally thrown stuff out of the sky at our column, including simulators. What if one of those simulators had gone down a tank hatch? Fortunately none did, but a lot of pretty nasty stuff did hit our vehicles."

The aviation commander is beginning to squirm a bit now because he realizes that this safety issue is serious and he wants to get out of this debate. He offered "Sir, if I can just get my log books back so I can continue training I am willing to forget the whole thing, including the simulators in the tents last night."

"I do not have the log books sir, but I will make an immediate inquiry and I will do my best to locate them or help the aviation commander recover them."

"You do that Captain Childers; NLT 0900 or you are in deep Kimchee."

They left. I summoned my First Sergeant and instructed him to collect the log books, put them all in a box or bag, take them to some location where they can be found; then call his counterpart in the Aviation Company and tell him where the log books are. No person to person hand off.

Make it into a training event. Have the scouts involved to provide overwatch and gather information on who recovers the log books, time of recovery, bumper number of vehicles involved, etc; and to infiltrate the airfield and place their CP under surveillance via field glasses and report arrival of log books; and to trail the vehicle that recovers the logbooks. I wanted evidence that they got their damn books back.

End of story. We got even for the aviation bombardment of our Troop road marches with flour bags, artillery simulators, eggs, water balloons, and something that smelled remarkably like re-processed beer.

Top of the Heap!
Carroll Childers, MG, ret

Making it to the top of the heap, commanding the famous 29th Infantry Division, was far more than meets the eye in my case. I was a renegade officer that operated well outside the norm compared to my peers and in comparison to what my seniors thought was acceptable by their model. I simply did not care. My rule was always to first prepare my troops for the ultimate experience of mankind and second to deal with whatever the fallout from that happened to be. So I decided and acted on my instincts, my concern for training soldiers, my golden rule to always set the example, and to place the opinion of my seniors about Carroll Childers last. I didn't publicly slap them in the face or blatantly disobey them; I did follow orders when given, but I also subscribed to the theory that forgiveness is less painful than permission.

My commissioned career began in the 29th Infantry Division in 1964 when I was commissioned as a second lieutenant by the Virginia National Guard Officer Candidate School (OCS). At that time, Major William J McCaddin was the Commandant of OCS. He was an Artillery Officer and wore a 2 ID patch on his right shoulder from having served in Korea. To my surprise, I was selected as the Distinguished Honor Graduate (DHG) of Class 6. I wasn't trying to be the DHG. I didn't even know there was such an honor until it was conferred on me. I was just working hard, doing my best, setting the example, and essentially doing those things that were my natural style and which became my trademark as life went on.

The bad part of being DHG is that there was an unpublished expectation, and unknown to me at the time, that the DHG would accept assignment in the OCS as a TAC officer for at least a few years. So when the system attempted to assign me to OCS, I declined. McCaddin's assistant, a Captain Frank Simmons, called me up to explain my "obligation" here. I simply told him, "I went to OCS to become a Platoon Leader; I have been commissioned in Cavalry, have a slot as a Platoon Leader in C Troop 183rd Cavalry Squadron in my home town, and that is what I want to do. I am not a screamer and have no desire to be a TAC Officer." Simmons was quite upset because he had the mission to rope me in and now he would have to go to McCaddin with bad news. I immediately called Captain Alvin York Bandy, my C Troop commander, and told him the story. "Don't worry about it LT," he said, "I will take care of those turkeys. You are my Platoon Leader and that's that."

I never heard anything else out of this but it was the first brick laid in a wall that would grow over the next 25 years. Major McCaddin later became the Adjutant General of Virginia. He went from being a LTC to being the Adjutant General, a two-star position. I discovered many years later that he was a guy who had a very good memory. Things rocked along for several years then there was a move at national level to re-organize the military. Out of this came the controversial decision to retire the 29th Infantry Division. At Annual Training 1966, the Division had a grand retirement parade. Then the politicians overcame this decision and the Division was revived by the stroke of a pen. In AT 1967, we had what the troops humorously referred to as "the second annual retirement" of the 29ID. Again, it was revived almost immediately. In January of 1968, I had to transfer to the Inactive Ready Reserve (IRR) in order to maintain my status as a Reservist when I took my first assignment in Vietnam as a VLAP Laboratory Representative. By the time I returned from Vietnam in the summer of 1968, the Division had conducted its third annual retirement ceremony..........and it did not get revived this time.

With the reorganization, the 28th ID, a Pennsylvania based Division, became the key major headquarters in the region and several Virginia units would be somehow attached to the 28th. That was a major blow to the ego of Virginia and Maryland because of the WWII fame of the 29ID, the Blue and Gray Division that led the assault on Omaha Beach, 6 June 1944. Under the 28th was the 223rd Cavalry Squadron, headquartered in Philadelphia. There was some powerful politics associated with this, but much of the decision came from a historic military organization known as the "First City C Troop Cavalry"; Philadelphia's Finest!! Because Virginia had an entire Cavalry Squadron under the 29th, a deal was struck to give Virginia one Cavalry Troop as part of the 223rd Cav line structure. So Richmond went from the headquarters of 183 CAV and 3 Line Troops to a single Line Troop; C Troop First Squadron, 223rd CAVALRY, 28th Infantry Division of the PANG.

From all of the officers in the 183rd, one James A. Bailey, Captain, was selected to command Virginia's only Cav Troop. I knew Bailey; he had been the commander of A-Troop before the Division retirement. He had spent some 10 years or so on active duty, wore the Ranger Tab and EIB, and really "talked the talk" about soldiering. A most outgoing officer with a propensity to dominate conversation with shallow charismatic declarations about most every endeavor imaginable, he was somehow simply a lot of fun to be around and plot mischief against our training adversaries. But beyond that, he would take the easy way out, avoid any personal implications, and basically practiced "do as I say do, not as I do." So he was fun, but one had to be cautious. But it took me a year or so to come to the above realizations.

Initially, he was very impressive and we young officers assigned to his command followed him with enthusiasm. He had a First Lieutenant as his Executive Officer, and 3 other Lieutenants as his Platoon Leaders. For the first two years with him, I was a Platoon Leader. Then he and wife became expectant parents for child number 6 and she told our XO he had to make a choice. I became the XO and this is when I really began to know the real Jim Bailey. I am sure that my membership in the CAV and my association with Jim Bailey had a lot to do with other's view of me and my potential to become a senior officer.

I had mentioned earlier that one had to be cautious in dealing with Bailey. Finally it came the day for Bailey to leave command of the CAV and for me to assume command. Such a move requires a change of command inventory and reconciliation of a lot of military property. The ordinary civilian cannot imagine what this entails. Property is basically maintained at the company level and from the standpoint of commanding officers, it is company commanders that sign for property. Organizationally Troops and Companies are equal and are commanded by Captains. As property rolls up to higher commands, Battalion through Division, commanders do not get involved in signing personally, but at the base level, company commanders sign on the dotted line and if something is lost or damaged negligently, that Captain pays personally. At higher levels of command, Warrant Officers typically are the unit property book officer and sign for all property and the management thereof,

C Troop had a secured supply room and internal to that was the even more secure arms vault. The arms vault contained all weapons plus those items of relatively high value and pilferability such as compasses and binoculars. We had a full time Supply NCO (SSG Leroy, Gene P.) and two M-Day assistants. One of the M-Day guys was Francis Bray, an Electrical Engineer from Dahlgren. Bray joined as an alternative to being drafted after discussions of alternatives with me at Dahlgren and he stayed there beyond his obligation until I left so he was loyal to me and had some insight into what was going on inside the supply room based on his intellect.

So came the day for Bailey and I to have what was called a joint inventory for turnover. We would together count and audit every accountable item and I would then sign for all items of inventory present and he would pay for all items not present for turnover. So we started with the motor pool counting vehicles and tool kits. Then we moved to the supply room counting mechanic coveralls, radios, blankets, boots, riot sticks, sleeping bags; you name it, we counted it. Only a few minor items thus far were out of balance. Then we moved into the arms vault to count sensitive items.

The count went good, all weapons present, until we got to the locker containing binoculars. There should have been 57 pairs; 40 full size and 17 smaller size. I have forgotten the model numbers now. A few did not have their leather cases and were just sitting there. Bailey is there trying to subtly rush me on to some other item but I am not only counting bare binoculars but also not assuming there is a pair of binoculars in each leather case. At the very back of the shelf, a binocular case is empty. This causes a recount.

There is a pair of binoculars missing. These are military issue items with ranging reticles and it was valued at $192.00; a lot of money in 1972. Bailey is laughing and joking about it, assuring me that this is all a mistake and that he will find them. I play total hard ball and show them as missing. This causes a formal investigation and he ultimately pays for them but for some reason, I get some of the blame by those up the chain who are looking for reason to give me another black mark.

The change of command inventory was done on a Friday. The following Tuesday, based on some feedback that Sgt. Bray had given me, I went down to the armory and surprised SSG Leroy by my presence.

"Good morning Sir, surprise to see you so soon since the inventory Friday."
"Yeah, well, always count on me to do the unexpected Sgt Leroy. Get the property book and let's go make a random check of property."

I could sense some discomfort at this suggestion. Leroy was somewhat like Bailey. I enjoyed him for his enthusiasm but did not totally trust him. We walked briskly back to the supply room and he went through the process of turning off the alarms and opening the doors and registering entry while making nervous small talk.

Sgt. Leroy needs a few words of explanation. He was a good soldier in terms of soldier participation. He loved to Cammy-up and go to the field and conduct training operations or to the live fire range. He loved being out there in it. He was also self-aggrandizing and far too susceptible to avoiding some truths to be responsible. I found out several years later that he had awarded himself both the Airborne qualification and the Ranger Tab.

I already knew what I wanted to count, but I started with some other items randomly, checking the count off on the property book. Then I came to the sleeping bags. The count came up 32 short!

"What the hell happened to my sleeping bags Sgt Leroy?"
"Sir, there is something you don't understand here……."
"No," I interrupted, "There is something that YOU don't understand here Sgt Leroy!!"
"Sir….."
"What happened to 32 sleeping bags since Friday?"
"Mr. Wheelmon loaned them to Captain Bailey for the inventory and he came to retrieve them Monday."
"Get Wheelmon on the phone right now. Tell him what I have found, put him on hold, then come to my office and get me and I will talk to Wheelmon."
"Sir, do you know who Mr. Wheelmon is?"

"Yes, I know who Mr. Wheelmon is; he is a CWO up in Boomtown who is the PBO for that Battalion up there and I don't know what business he has here in my Troop supply room. Get him on the horn!" And I turned and walked to my office.

So who is Mr. Wheelmon? He was a full time Guardsman up in Boomtown Va. Who had been a member of the Guard since the end of WWII, had taken the Warrant Officer route because Warrants have a certain un-touchable aura of invincibility as they linger somewhere between senior NCOs and Officers. Typically they are associated with logistics or maintenance functions. Wheelmon was the "king of logistics" in Virginia and was known widely and covertly as the go-to-guy if you are in trouble with property. Because he had been dealing in property, at that time, for some 25 years, his reputation around the state and even across state lines was considerable. Later in my career I would find that he operated as a silent (secret) partner in an Army Surplus store in Boomtown, clearly a conflict of interest. I would also find that he had several storage containers behind the Armory which only he had access to and there was always speculation as to exactly what these contained and who owned what was inside.

"Good morning Mr. Wheelmon," I opened.

"Good morning Captain Childers. Congratulations for the assumption of command," he replied as a probable delaying tactic.

"Mr. Wheelmon, it appears that you came to my supply room yesterday and removed 32 sleeping bags from my shelf; now, I don't care what the circumstance was of those 32 sleeping bags getting on my shelf. That is your problem. My problem is that I am 32 bags short and I have a SGT here that will testify that you took them away."

"Captain Childers, if you will do your homework fully and ask your SNCO he will tell you that I presented him with a hand receipt for those 32 bags and that I merely reclaimed bags that I had loaned him."

"Well Mr. Wheelmon, I have not been through an IG before so I don't know where the IG may lay the blame; on you as a knowing participant or on Bailey for misleading you as to why he needed to have the bags on the shelf but I will tell you this; If my bags are not back in my supply room by Friday, I will initiate a IG on Monday and name you as a possible willing/knowing participant. Have a good day."

I was told much later that he stood up, grabbed the phone and destroyed it by slamming it down on the concrete floor at Boomtown. He then marched back and forth ranting and cursing for 10 minutes. In later years, he and I became much closer after a few intervening encounters. He was actually a very valuable guy to have on the team but he did require a strong hand to ensure that he did not take too many liberties with his assignment in life. He passed away several years ago but was a strong supporter of the Guard and the National Guard Association until he died.

Mr. Wheelmon returned the 32 bags, Bailey was charged with the missing binoculars and a few other items that were short, and my command resumed. I was not on good terms now with Mr. Wheelmon and his criticism of me penetrated upper circles as valid because of who he was.

I had the great pleasure to command the Cavalry Troop for 5 years. I would still be there if I had not been moved by others. The Troop was conducting the weekend drill. I think it was perhaps the January drill. Typically we did not go to Ft Pickett for weekend drill during December, January, and February because the weather was simply not predictable enough to support that. What has the weather got to do with it? Well, to begin with, we convoyed down by open vehicle on a Friday night, spent the night in the field, and woke up early Saturday morning and trained until Sunday noon. Then it was vehicles to the wash rack, turn all back in that we had checked out from Ft Pickett, and convoy back to Richmond. Very cold weather can make this regime most uncomfortable. Most of the time we were right; sometimes it would be colder in November than it might later be in January but because we had to coordinate with using Ft Pickett, we had to play the hand issued. Most of the time we were ok.

So back to the story line, we were conducting weekend drill at the armory; I got an unannounced visit from COL Bradshaw on Sunday afternoon about 1200. This was most unusual. I knew that he had not driven from Staunton for some small talk, especially with me. The armory was a shared facility, ie; other units had space there and we shared and coordinated the drill floor for formations. After greetings and offer of coffee, Bradshaw said "Let's wander around and look the armory over."

After going down every hallway in the building and walking around the drill floor, I said "Col Bradshaw is there something particular you are looking for; perhaps I can help you find it?"

"Well, I was hoping to find an empty class room or someplace we could talk privately and your office is surrounded by partitions and not too private."
"It is a nice afternoon, how about we walk out to the parking lot and talk?"
"Good idea. Yes, that would be fine." So we walk out about 50 yards into the rear parking lot that is sparsely enough occupied to ensure privacy.
"How long you been here Captain Childers?"
"Going on five years, Sir."
"That's entirely too long for Company Command."
"This is not a Company Sir, this is a Troop and one can never command a Cav Troop too long."
"Aw Bull shoot Childers; that's part of your problem. You think you are sumpthin special and these bunch of hooligans you call Troopers are sumpthin special and you need to find out about the rest of the army...........so I am moving you. Got any problem with that Captain?"
"Where are you moving me to Sir?"
"Doesn't matter. You got a problem with moving? That's the question!"
"So long as you know that I am not asking to be moved. I am perfectly happy here."
"Asking is not in the conversation Captain Childers; I am telling you I am moving you."
"Where to? What job?"

"What I want to say," he started "is simple. You have been here now for 5 years commanding this Troop. In spite of the little trouble we had at Ft Bragg on your letter writing, you have done a good job as a commander. But 5 years is way too long. You need to move and I need an Assistant Intelligence Officer and I think you are the right person for the job."

I knew what the Asst S-2 function was and I knew who the S-2 was and I knew that I would effectively be in charge so I replied, "OK. When do you want me to start?"

He was obviously shocked. He expected me to put up a defense and delaying action. He almost turned red again but recomposed himself to appear in charge and successful. "Why don't you make one more weekend drill with the Troop. That will give you time to do the change of command inventory and say good bye to the men."

"Yes sir. Thank you for the opportunity to move to a new experience. You probably know that I would never have asked for a move. I am always happy where ever I am but also that I will do what I am told.............if I believe in it. So this is good for both of us."

"I have not forgotten the job you and the Troop did at Bragg. That told me that you can do the S-2 job and we need that section to be energized some so we are looking forward to you coming on board." With that, he saluted and bid me good bye. He never again mentioned the hair letter nor the events that caused me to remain in command.

So it was goodbye Cavalry, Hello Infantry.

Infantry was not totally new for me, after all, within a Platoon of Cavalry there was a Rifle Section (MOS 11-B) that traveled in an M-113 as well as a Scout Section. The Scouts normally moved via gun jeep but were essentially infantry soldiers with a special mission and a different MOS (19-D). So I had done my share of dismounted operations but now I was going to be INFANTRY. I was to be the Assistant Intelligence Officer (Asst S-2) for the 116th Infantry Brigade.

Thundering Herd
Rick Meador

As if the wind had blown away all traces of those souls,
No longer heard are the bellowing sounds aimed at distant knolls.

From hidden trails to APC's securing the nightly stay,
Standing here I still can see the experts stalk their prey.

Great leaders create dedicated troops who execute precision skills,

Neither Agony or Misery could pose as formidable hills.

Time has yielded to the past but veterans minds still recall,
The training, meals, longs days and nights met with strong resolve.

Listen closely, my friend, and you may hear the rumble of the "Thundering Herd"

Sleep well Virginia, all is fine...tis the sound of the 223rd.

Dedicated to the Great Officers and Troopers of the 223rd Armored CAV from one of its humble sons. Rick Meador, August 2019

Woodsey Ragland's Wash Point initiation
SPC-5 Mike Thomas

"I'm just back from basic training...my first time at Pickett and It was our turn to wash down our tank for turn in. Tank covered in red clay. Everybody is in rain suits -- except me -- and wading around in the water built up on the wash rack...the drains couldn't keep up.

Some dude says 'Ragland, come over here.' I go over and he hands off the hose to me...it felt like a fire hose...so much pressure! You had to man handle that thing. Then, the dude walks off and leaves me. 'Hey man, come back here. Don't leave me with this hose all by myself!'

"You couldn't just put that thing down, you know!"

When Woodsey says that one has to have had some experience handling a water hose with an operating pressure of perhaps 150 PSI, and a high flow rate (in gallons per minute). As is said among scientists, for every action there is an equal and opposite reaction. A water hose is a crude jet engine. Water goes one direction which means the hose must go the other direction and if someone drops the hose they better either run out of its bullwhip like whipping range or they must dive on it like smothering a grenade in combat. Of course if you lose control of the hose, best to go find the shut off valve and soothe it.

C Troop spent a considerable amount of time on the Wash Rack. Unfortunately the picture presented in support of this tale is the only image of tracked vehicles on the wash rack and sometimes when we were issued a rack, we drew the fourth class hose systems due to competition for priority of the number one racks.

Just another crow that C Troop often ate. Or was it Possum?

Why I was never selected as the Adjutant General of the Virginia National Guard
Carroll Childers, MG, ret

My career in the Army National Guard spanned almost 44 years, beginning in October of 1955 with enlistment and ending in August 1999 with retirement. Over most of this time I was working as an Engineer for the Department of Navy (28 years at NSWC Dahlgren and 10 years at Quantico Marine Corps Base). As a Federal employee there were occasions when I was deployed Out of Continental United States (OCONUS) into war zones as a civilian and when this occurred I would have to be transferred into the Inactive Guard to account for my absence. These OCONUS times totaled about 14 months so this amount of time must be subtracted from my total National Guard service time.

Overall my parallel career (National Guard and Engineer) was very satisfying and successful. My rating as a Civil Servant was GM-15 and my grade as a National Guard Officer was 0-8, Major General (2-star), commanding the 29th Infantry Division in Virginia (with some units in MD, NJ, MA, and CN).

Some may be curious about why I was never selected as the Adjutant General of the Virginia National Guard. That seems like a natural curiosity for after all, I had accomplished a number of outstanding command performances in my career. I had started my career by attending the State Officer Candidate School (OCS) and graduated as Distinguished Honor Graduate in Armor. I stayed in Armored Cavalry from 1964 until 1976 having commanded the Cavalry Troop for over 4 years before being reassigned to the 116th Infantry Regiment. Remaining in the 116th until I relocated to attend the in-residence course (64-65) of the Army War College (AWC). I also squeezed in several staff and command positions in the 116th and also attended the US Army Ranger Course, graduating as the

Distinguished Honor Graduate at age 42 years and 8 months.

Upon graduating AWC, I was assigned a second 0-5 command, awaiting a position in the newly organized 29th Light Infantry Division. Finally, in 1989 the Division Commander retired which caused a ripple effect down the Chain Of Command and I was selected to command the 2nd Brigade and promoted to 0-6, Colonel. At this point I am one grade from being a "Flag Officer." This means that my appointment is so important as to require approval by Congress.

This grade is also the launch pad to be considered for the rare rank of Major General at the state level. Typically, a State will have not more than 2 officers at the 0-8 level; The Adjutant General (TAG), and if the State also contains a Division sized unit, it's commander is also a 0-8 position. But at the time I became a Brigade Commander, my long range sights were on the position of Division Commander, not TAG.

TAG is an appointed position which serves at the pleasure of the Governor in every instance but South Carolina where the TAG is a political position which is elected by the populous.

Now, allow me to insert a slight deviation. There was a time later in my career as a Division Commander wherein a new Governor was elected and a TAG appointment was a foregone conclusion and therefore I did make an effort to gain that appointment more because I felt an obligation to make myself available than because I truly wanted that job more than any other.

As it turns out, the new Governor was one Jim Gilmore who took office in 1998 for four years. I submitted my resume and was awarded an interview. It was a very positive interview and I left with a feeling of success but not of confidence. A 0-6, an Engineer Branch Officer was selected as his TAG.

For some reason I was surprised when I received an invitation to the installation of the new Adjutant General (AG) at the south portico of the Governor's Mansion in Richmond.

It is always a good idea to accept such invitations and I attended with the greatest good wishes for the new AG. I stood at the back of the assembled group while the new Governor attached the rank epaulettes on the shoulder of the new AG. I noted that the Governor gazed my direction and it was somehow obvious that he picked me out of the crowd for some reason. When he had completed his remarks and the formality was done, he set an immediate course to intersect me. Within a few seconds he was shaking my hand and welcoming me to the ceremony and thanking me for attending.

He went on to share something with me that he certainly did not have to do. "I just want you to know that you were the better of all those I interviewed for AG, but, there are politics in these kinds of positions and I had obligated myself to appoint (name redacted) to the position as a favor to my Attorney General who you may know to be a personal friend and neighbor to the appointee. But for this truth, you would be TAG. Thank you for your many accomplishments and I hope you will continue to serve us."

I continued to command the Division with distinction and accomplishment until retirement in 1999. At that time there was a regulation which referred to a Mandatory Retirement Date (MRD) which was linked to age. At the time I passed command of the Division to one of my ADCs, I was only one year away from MRD. As the post retirement years moved along, I learned a lot more about the availability of other opportunities to serve than I knew when I retired.

In 2004, my life long partner discovered she had Stage 4 breast cancer. This began an entirely different life style for the two of us. Surgery and Chemotherapy caused remission of the cancer but cancer is a terrible misfortune as it may appear to be cured but re-appear in some other form or organ at most anytime.

In 2009 there was an election in which the Governor of Virginia was changed. This prompted the plan to appoint a new AG. A longtime friend of mine named Tom Berozowski called me in early February, as he had done many years before when I shot for TAG under Gilmore. Tom was tasked to present the new Governor with a list of candidates for various appointments and Tom asked me if I would consider being TAG.

"Tom I can tell you that that would be a great honor but tell me more as I thought the current AG was doing a good job?"

"Carrolll I am not at liberty to comment on the current AG, only that he will not be reappointed."

"That is fair enough, Tom; I cannot say yes or no right now. My wife may be experiencing a resurgence of cancer; we do not know for sure. But I will have to talk it over with her as I would not want to become TAG one month and have to resign shortly thereafter."

Dayle and I talked about the opportunity and I surely would like to have been able to get back in uniform. It would have been somewhat unique if a graduate of VANG OCS grew up to be TAG! But it was not to be. Dayle survived just over 4 more years after we made that decision and much of the time I was engaged with her in treatment and attention and making her life as pleasant as possible.

I called Tom back to give him our decision. Not being able to accept this opportunity was one of the most difficult choices of my life but the important thing was to support Dayle in a time of great trial and to do all I could to enable the selection of the very best officer as TAG. I asked Tom if he had considered MG Daniel E. "Chip" Long.

"Who is Chip Long?" was his reply.

So I gave Tom the update on Chip: Had been my Brigade XO and took Command when I was promoted to BG. He earned the Ranger Tab and EIB (my requirements for him to be my XO) He then became ADC(M) then Division CG. Deployed to Bosnia. Deployed to Hurricane Katrina and to SW Border. Deployed to Iraq to be senior Contracting Officer of rebuild program. Currently CG of a Joint Command at Ft Monroe.

"Tom, Chip has been out of sight/out of mind now for several years but he is about to retire from his current assignment at Ft Monroe and he is the best person available for consideration as TAG."

I gave Tom the necessary contact data for MG Long. "You call him up and I will alert him you will be calling."

I called up Chip and told him the story. He replied "Well, MG Castles (a lifelong mentor to us both) would want one of us to do this so if it is not you then I will apply. I had intended to do some serious fishing but, duty first"

Chip was selected out of several applicants and was in my opinion the most successful TAG in Virginia history.

Life in the Cavalry
COL Tom Redfern

I am going to make a concerted attempt to write down the many memories I have in a chronological order. Keeping in mind that age is beginning to take its toll on my memory, here goes:

In the spring of 1971 as I was getting ready to graduate from college, I received a notice that I needed to appear for an induction physical – my draft number was 52. After I tried everything in my power unsuccessfully to convince the medical station that I was not fit to serve from having a chipped bone in my vertebra years earlier to any other medical problem I had ever experienced, I immediately reached out to LT Bugg of C Troop who is a friend and he suggested I go to Dove Street and get my name on the waiting list. I went there and met with COL Merkel to have my name placed on the list. As I registered, he was laughing because the list was at least 10 pages long. I don't know what happened but the week after graduating, I received two letters, one from the Department of Defense and the other from the Virginia National Guard. Both had "greetings" in their salutations. Without hesitation, I accepted the National Guard letter. Low and behold, I was assigned to C Troop 223rd and was informed to report to the orderly room where I met with Sergeant LP Hening and First Sergeant Donnie Knapp. LP placed me in the 3rd platoon scout section.

I shipped off to basic in the fall of 1971 to Ft Knox. If you went through basic around that time, once you graduated from basic and were assigned to AIT, the First Sergeant would stand in front of formation and reel off names and assigned locations for AIT (Benning, Knox, Polk, etc). My name was the last one called and I was being shipped to Ft Polk – the only soldier in this basic training company going there. I was devastated; LP had lied to me because I was supposed to be a scout and stay at Knox. Once the first sergeant dismissed the formation, I immediately went to the closest coin phone and called the Adjutant General's office (McCaddin) and began to jump up and down complaining that I was not supposed to go to Polk but was supposed to stay at Knox. Something happened because my Senior Drill Sergeant informed me that I would be assigned to a scout AIT unit at Knox but would have to stay in the basic barracks another week to join the next rotation.

After graduating, I reported back to C Troop. The first weekend I drilled as a PFC, we were bivouacking at Ft. Picket and I had to sleep in a pup tent with my platoon sergeant, Johnny Thompson. When we woke up the next morning, there was snow on the ground. The next drill weekend was at Dove Street when LT Bugg pulled me aside and said I needed to go to OCS immediately and I had to get my name on the roster to join the upcoming class (class 15). But, in order to be considered, I needed my commander's recommendation. LT Bugg arranged a meeting between him, CPT Childers, and me in the supply room to discuss going to OCS. Childers was sitting on the supply bench just in front of the vault and the conversation was going smooth and casual with them telling me what to expect and how the program was run. Then, all of a sudden, Childers jumps off the bench and starts yelling at me asking if I could take it. Scared the hell out of me while Bugg was in the background laughing his butt off. I must have impressed him because I was at Ft Pendleton for my first AT camp in OCS.

My first drill back after being commissioned was at Dove Street. LT Bugg had retired and the 3rd Platoon Leader position was held open for me. I came into drill Friday night all spit shined, uniform pressed and starched to perfection, a perfect example of a "green" lieutenant. After formation, the rifle section sergeant Deats came up to me and asked if he could speak with me outside in the gravel parking lot. Of course, I said "yes." Once outside, he got in my face puffing his cigar, similar to the flight commander in "Top Gun", and said "Sir, we will make you or break you," and he proceeded to kick gavel on my highly spit shined Corcoran boots. Believe it or not, those words stuck with me throughout my entire military career.

The first memory I have of being a platoon leader in C Troop was during a summer camp at Pickett. We had been on a daytime road march and one of my tanks blew a track. Naturally, I called SGT Harver and the maintenance section who were quick to respond. We were all standing around watching them do their thing when Harver told me I didn't need to be there so I headed back to the company area. It was dinner time and everyone was in the mess hall talking shop and reminiscing about the day's activities. As was the standard, the officers had a table at the back of the chow hall where Childers, Pounds, Deverell, and I were seated. Nonchalantly, Childers asked me where my disabled tank was located to which I answered it was having the track placed back. He immediately started chewing my butt, telling me that I should be out there with the men and that he was going to carry me unsatisfactory for the weekend. After Pounds convinced him to not do that, I immediately hauled my butt out to the site where Harver asked me why I was there. After I told him what Chiders had said, he started laughing.

At some point early in my days with C Troop, Duke Karnes had returned to Richmond after leaving North Carolina where he had moved for his job. It was a Sunday after drill at the Dove Street armory and some of us had gathered for a beverage or two to discuss the weekend's training. If I remember correctly, the "club" room was in one of the rooms across the hall from the orderly room.

Duke called me aside and said, "You officers aren't worth a ___!." The Duke and I have been close friends ever since.

I am sure this one has been told but I want to make sure I substantiate the tale from my perspective. At another summer camp at Pickett, Childers and the officer corps were up to no good. We had a few beverages that evening. I can't remember whose room we focused our attention on but we had a grenade simulator that we wrapped with a water-soaked roll of toilet paper. That seemed to be the thing to do in those days. We pulled the pin and threw it into the room we had designated as the objective. The explosion not only stuccoed the room, but the windows, wall hangings, and anything else that was in the vicinity blew out and broke. We quickly made our exit from the building, not through the doorways but crawled out of some windows onto the middle roof and worked our way down, then jumped into Childers' truck. He was driving and I am not sure who was in the cab with him – probably Pounds and/or Deverell because Terry, Call, and I were in the bed with the beverage container. I think we had issues with Mrs. Banks in facilities when we tried to check out Sunday afternoon. (Editor's note: a portion of this recollection has omitted to protect the innocent.)

One story that has been related by others in this journal was a particular night road march in which SSG Talbott's tank nearly ran off a bridge and took out part of the railing. It was very dark and the tank trail was extremely dusty with the 3rd Platoon pulling up the rear — which I believe Childers always planned for us for some reason. To make matters more complicated, the march was conducted in the southern portion of Pickett where we had seldom trained. Davoud and I were up near the front of the platoon when I hear Talbot come across the radio saying he had a real problem and to stop the convoy. I immediately radioed Childers and went back to Talbot only to find his crew had dismounted the tank because they were hanging off the side nearly tipping over into the Nottaway River. It seems like everyone in the troop gathered at the scene with Childers pulling Harver in who said he would take care of it. Harver coolly climbed in the driver's hatch and backed the tank up out of harm's way. I am not real clear of what happened next but I have been told that Childers proceeded to chew my butt up one side and down the other. I am sure LP enjoyed dealing with the required report of survey paperwork.

Again, another tale that others have told is the time Talbott (does he sound like a recurring theme?) stuck his tank in the swamp. This may have been at the same summer camp as above because I believe this happened in the southern portion of Pickett as well. Talbot radioed me that he was stuck. I went back to his location only to find his tank had sunk up to the deck; you could literally walk off the ground onto the tank's deck. There was no way for the platoon to recover the tank so, once again, I radioed Childers who came down to the location. After analyzing the situation, he radioed Harver to come down with the M88. After the M88 arrived, Harver connected the boom cable to the tank and attempted to pull it out. The tank did not budge because of the suction created by the tank's belly being completely submerged. Childers took an entrenching tool and began digging a hole to allow air to get up under the tank to relieve the suction. After learning my lesson about the tank that had thrown a track, I immediately picked up an entrenching tool and began digging alongside Childers. In the meantime, Harver had directed another tank to pull up in front of the M88 and he connected that tank with a tow bar to it so we were pulling in tandem. Eventually, the tank popped up out of the bog. I don't believe there was any damage to anything but there were 2 officers covered in mud up to our ears and Talbott's crew probably spent all day on the wash rack.

At another camp at Pickett, we were on maneuvers conducting a recon. I was interested to see how my tank section was doing as it pertained to cover and maneuver but more particularly with being in a hull defilade position. Davoud and I went out in front with the scouts and the distance from the tanks was a good 500 meters or more with an open field between us. I asked Davoud to stop the quarter ton in a covered area so I could get a good look back. When I did look back, all I could see were the turrets exposed because the crews had picked out perfect locations behind terrain features that allowed them to be in the hull defilade mode. Actually, I had to use my binoculars to spot them. As I began to scan the tanks, I noticed Jeff Russell, one of my tank commanders, in his position on the tank with his binoculars looking at me. I waved back to him and he, in turn, picked up a martini glass as if to toast the maneuvers. All of this through binoculars. I didn't ask any questions.

One final note, as well as a tribute to COL (Ret) Redfern. Tom spent his time in all TOE positions in the Troop and did them all well. As I recall he followed Deverell in Command then was reassigned elsewhere and was replaced by a C Troop commander who was not experienced sufficiently to lead, train, and set the example for the complexity of an Armored Cavalry unit. This interim commander was reassigned after perhaps a year and Redfern was recalled to take command for a second time. He was in command when a reorganization once again caused the casing of the colors for the final time on Armored Cavalry. The unit was re-designated ASA 0-5 command within the 29th ID and continues the traditional task of reconnaissance Cavalry, non-armored.

Flood Duty
SPC Alan Hughes

Our Troop, amongst many National Guard units were activated to assist the City of Richmond during James River floods caused by hurricanes Camille and Agnes. Most of the Guardsmen were called on to provide supplemental security manpower.

Coordinating activities of the units was done through Virginia's Adjutant General's command. A tragic incident at Kent State University prompted orders for units such as ours to be posted in effected areas with weapons but without ammunition. Company commanders took exception to this order since our people were principally assigned to security details, helping prevent looting from businesses and residences which were under mandatory evacuation orders due to the severe flooding.

A compromise was reached. Guardsmen serving guard duty assignments would be issued an M-16 with a magazine holding a three rounds of live ammo. However, the magazine was to be taped to the rifle strap or to the soldier's web gear...not loaded in the rifle.

A Long Night

I boarded the duce and a half (a 2.5 ton capacity, high water, three axle cargo truck) at our Dove Street armory along with a truck load of other Troopers. We were dropped off at assigned points in pairs to pull a four hour shift. But, when it was my turn to be tagged to hop off the truck, it was just me...no partner. It's the middle of the night, like two A.M., no moon, overcast and extremely dark. The nearest light was forty yards away, a single bulb on the side of a warehouse. I walked around to become familiar with the area. I have no idea where I am but it's an industrial area. Two long warehouses were side by side with a road between them. I walked twenty yards in one direction and stepped into water…the James River.

A white car kept passing back and forth on the main drag, appearing to be checking out the area. I could not see the driver or passengers. After the fourth or fifth pass, I'm getting worried…this is not good. Out of nowhere a young boy approached and said, "Hey mister, you got bullets for that rifle?" Holy smokes! I tore the tape off the magazine, showed the boy the bullets and slapped the magazine into my rifle. "Boy, you go tell the men you are with that I'm locked and loaded!"

I began to wonder just what's in those warehouses? Something those guys in the white car would love to have, no doubt. How long is it until daylight? When will the guys come to back to get me? Will I be attacked, shot at by someone who wants whatever it is in this warehouse?

I decide to take up a position out of the light but close enough to where I could keep watch on the road and the warehouse approaches. I find a concrete window sill and sit, watch and wait. Come on, show me some sunshine. Hours passed. Thankfully, the suspicious car had stopped circling. Daylight was breaking and there it was, at last…my ride! Never been so glad to see a duce and a half!

Tank Commander
SFC Johnny Thompson

My job was to drive Johnny Thompson's tank. From the beginning, we did not get along very well. You might say that we "didn't see eye to eye."

Hornet attack!

Driving down one of the many steep slopes that you encounter frequently in the unique terrain that is Fort Pickett, our tank threw a track. Our crew dismounted. I quickly discerned that Tank Commander, Sgt. First Class Johnny Thompson was not pleased judging by the stream of nasty words directed at me and my lousy driving that resulted in this breakdown. Losing a track is not easily recovered from...we were out of commission — for the day.

Suddenly, Sgt. Thompson stopped yelling at me and started yelling in pain. We had run over a nest of ground hornets and they were attacking him! These creatures are as big as your index finger. In a coordinated attack, they were hitting him hard in the face. I grab him and we hustle away from the nest, swatting away at the hornets as we ran. Many yards away I am battling a few persistent, angry hornets. Sgt. Thompson's face was swelling quickly, appearing to me approaching the size of a basketball. A call for help and Johnny is taken away to get medical treatment for multiple stings. I'm okay, I think, so I did not go to the medical facility at first. Later I realized that was a mistake when my fingers were swelling, growing together. Amazingly, a shot of meds and soon I was mending as was Thompson.

Oddly, it took a hornet attack and our mutual suffering to help mend our fences. Next time we had a weekend drill, Johnny gave me a big hug.

Poker Anyone?
SPC Allan Hughes

C-Troop had a bunch dedicated poker players. We would get a game going whenever possible, always after duty hours, of course.

Sometimes we played in the club but If were in the barracks area there was a mess hall available - and the mess steward wasn't around to run us out - we'd deal the tickets at one of the large tables. One of our cooks was a regular player and he knew where the cheese slices and loaf bread were kept.

Nice hand!

This story is from a drill weekend at Pickett. There were six or seven players in a hand of standard seven card stud, nothing wild. As the deal progressed, the pot was building as most early bets were raised and re-raised. After the sixth card, only two players remained, myself and Davoud. At this point, I'm was looking at three 3's. Last card dealt down...it was the fourth 3! I bet and Davoud hits back. Thinking Davoud might have a full house at best, I raised only to be quickly challenged by another raise from Davoud; the last one allowed. It is now time to show the hand.

"Not good enough, Roddy!" I said proudly and confidently lays out his four beautiful 3's.

Roddy grins and replies, "Not good enough, Alan!" laying down his hand...four 6's.

After nearly fifty years, we still reminisce about that hand.

Soldier! You Owe the Army $87,000!
As told to Mike Thomas by Bill Talbott

One C-Troop's First Sergeants, Bill Talbott, began his military career in 1956 by joining the Air Force. He was trained as a radar operator and served at Lackland, Keesler, French Morroco and North Africa. He left the Air Force in 1961 to get married and settled down. In 1963, Bill signed up in the Virginia Army National Guard, 29th Infantry Division, C Troop, 1-183rd Armored Cavalry Squadron, under a "Try One (year)" program as a PFC radio operator.

The 29 ID and all of its subordinate commands except C Troop was retired in February 1968 and the single Troop was actually a part of the 1-223rd Armored Cavalry Squadron, part of the 28th ID of PA. Bill transitioned via On The Job Training (OJT) to become a tank commander and then platoon Platoon Sergeant. Later, Captain Redfern asked Talbott to take on the First Sargent role just for an upcoming AT. That temporary assignment ended up lasting until 1982 when the Troop was converted to an infantry outfit. Bill Talbott was the last to hold the First Sergeant post in C-Troop.

Back in his Tank Commander days, Bill and his crew had several incidents, some of which are noted in this publication. Some say that every photo we have of a tank that's thrown a track or otherwise disabled is a picture of Talbott's tank. "Yeah, that mess must be Bill's tank."

Anyone who's had the pleasure of being in the middle of a convoy when C-Troop pulled out of the Pickett tank park with over 20 full tracked vehicles and about the same number of wheeled vehicles headed to the field in July knows about the dust. One particular year, we arrived for AT in the midst of an extensive drought and the dust was incredible...a footprint was an inch deep the ground was so dry. Our CO brought huge, red, cowboy bandanas to help filter out the dust stirred up by the tracks. Hanging back from the vehicle in front of you was a necessity, lest everyone choked to death on the plume of fine dust cloud, not to mention the distinct possibility of running into the vehicle ahead of you.

That's what happened. Going around a sharp curve and then down into a bottom, Bill's driver could barely see a few feet ahead. Even at a very slow speed, given the near-zero visibility the tank-to-tank collision was serious. No one on either crew was hurt but the lead tank had substantial damage and had to be towed from the field. The tank's final drive was damaged to the point that it could not be repaired.

As you'd expect, the folks who "own" the tanks at Pickett require thorough reporting and investigation of any vehicle accidents and this one was a doozie. Long after the investigation was complete, months after that AT was over, a letter came to Bill's house.

In the letter, the Camp Pickett "powers-that-be" said they had determined that Bill, the Tank Commander of the following vehicle which collided with the vehicle he was following, was negligent and therefore personally liable for reimbursement of the cost of replacing the drive system. (As in auto accidents, the person that runs into the back end of another car is almost always judged to be at fault.) Included with the letter was in invoice made out to Bill Talbott in the amount of $87,000.

$87,000 is a lot of dough these days but back then it was a fortune! The Army's invoice was no joke...they wanted the money!

After several months of appeals and further discussions — and many sleepless nights for Bill — the Pickett authorities relented and wrote off the event as a routine training accident with no liability.

Sometimes we forget that when we went to the field, many millions of dollars of hardware went with us.

Another Invoice for Talbott
As told to Mike Thomas by Bill Talbott

C-Troop traveled to Fort Bragg for the 1982 Annual Training. The Troop drew several WWII vintage buildings arranged in a "company area" configuration...three barracks, a mess hall, and an orderly room/supply area combo complete with a secure weapons storage area. Pretty convenient set up.

In the middle of the second week of AT and after a long, hot day of small arms training and qualification sessions, First Sergeant Talbott ordered the weapons used stored for the night. "We'll clean everything in the morning...just secure them in the armory for now."

The next morning, a detail of Troopers reported to Supply Sergeant and Administrative Supply Technician (AST), Gary Brooks to begin cleaning a stack of weapons including lots of M-16's and both of the Troop's M-203 grenade launchers.

One trooper, picked up a M-203 and said "sarge, how do you clean one of these?" Brooks took the weapon from the trooper to show him how. "Son, this is how it's done..." rapping the butt of the weapon on a table. Luckily the barrel was pointed straight up! It discharged. Someone had left a smoke round in the weapon during the hurried turn the night before.

The round launched up through the ceiling. Smoke billowed out of the ceiling, filling the supply room. The Troopers cleared out fast and reported to Talbott who alerted the Ft. Bragg fire department who came a-running, sirens wailing and lights flashing.

The fire crew pulled down a section of the supply room ceiling where the round had punched through. They confirmed that nothing was burning...just lots of smoke.

Having to call the fire department was bad enough but leaving a live round in a stored weapon was the cardinal sin.

Adding to Bill's ongoing angst over the incident was the Army's invoice in the amount of $246 to repair the supply room ceiling.

C-Troop pulled out of Bragg a couple of days later headed for Richmond. Bill thinks the invoice could be still outstanding.

Golf Anyone?
SSGT Roddy Davoud

One Saturday at Fort A.P. Hill, the CO had us scheduled for a long duration night road march. (These after-dark marches must have been a favorite of whomever it was that set the training schedule for armor units because we sure seemed to participate in more than our fair share.)

Our orders were that all personnel were to report to their vehicles in the tank park promptly at 1700 hours with the order to move out scheduled to come at 2100 hours...darkness hours in the summertime.

Having a few free hours before the 1700 assembly, we, young Troopers Davoud and Hughes, decided that there was plenty of time for some golf at the Four Winds Golf Course, only 30 minutes south of A.P. Hill.

After finishing 18 holes, with almost an hour to go to formation, we return to the car, get our fatigues out of the back seat and change in the club house. We return to the car, opening the trunk to store our clubs and golf attire and shut the trunk. Oops! And this was the BIG oops! The car's keys are in the pocket of the golf shorts just packed away in the now locked trunk!

Thank goodness that the interior of the car was unlocked...we have to get into the trunk somehow...time is running out! Being late back to post was not an option!

Try as we could, we were unable to remove any portion of the back seat. We are in serious trouble, deep trouble if we are not on site and ready to go on time. I was Johnny Thompson's tank driver...he won't be happy if his driver is among the missing. Roddy, a scout section leader was sweating bullets about what the CO was going to do to us. Panic time, deep trouble!

Finally, a desperation thought occurred. I ram my fist through the material covering the rear speakers on top of the back window shelf. I fished my arm through the speaker hole and got my hands on the suitcase and managed to open the latch. I grabbed a bunch of clothes, then another bunch and pulled them through the hole.

Roddy says, "I have the keys! Let's hustle!"

My reply, "Hey, Roddy, do we have time to grab a beer on the way back?" This is no doubt the second most desperate of the tales.

Lead, Follow, or get out of the way
Carroll Childers, MG, ret

Lead, Follow, or get out of the way. Crude but sometimes effective, depending upon how, when, on whom, by whom, the scenario, and what the motive is for its use. Undoubtedly it is a phrase implicating the employment of leadership.

There are many definitions of leadership. Mine is, "whatever works', but with this in mind, there are some proven guidelines that will assist a leader in making their mark on the led.

There are areas in which a leader should have low tolerance:
- Breach of integrity (this is first because integrity is the soul of the Army).
- Failure to follow instructions.
- Wasting time and energy – yours, mine, others.
- Not asking for, or giving, help.
- Being miserable on my time.
- Taking credit without giving credit.
- Shifting responsibility.
- Lack of style/class.
- Sexual, ethic, racial, and behavioral stupidity.

Some words to live by:
- Call people by rank and name.
- Be specific when criticizing; and be private.
- Cultivate order.
- Listen-to-speak should be as 4-to-1.
- Enjoy what you do.
- Never stop learning; never stop teaching.
- Overcome fear internally.
- Be a hands-on company-grade leader.
- Compassion is not a sign of weakness.

- Curiosity did not kill the cat.
- Compliment good work; correct bad work.
- Think before acting, especially if it is a scenario that is time tolerant.
- Know your job, your mission, your capability, and the enemy.

Motivate and lead by example:
- Build for the future – train and counsel.
- Acknowledge subordinates.
- No backstabbing.
- Involve yourself at the right level and at the right intensity.
- Play the course, not the competition.
- Perform every assignment as if it is the last one you will ever have.
- Prioritize what you assign and what you accept
- Always understand (and make understood) timelines and outcomes (intent).
- Share good news.
- If you don't enjoy every moment, you are in the wrong job.

Stuff that may stress the last bullet above:
- Seek, assume, accept, and monitor responsibility.
- Know not only the latest doctrine and TTPs, but things that were once used.
- Take chances, admit mistakes, look back at only the lesson.
- Don't accept the status quo. Many things are done because no one ever asked.
- Find your lane. Run hard. Coming in second means you are the first loser.
- Understand and partake in personnel matters.
- Understand and partake in logistics matters.

- Every action can produce unintended consequences and if you overlook the last two bullets above, you can forget the rest of the advice and pick the color of the hand-basket that you are going to hell in.

Then, don't forget how important a sense of humor is to the balance of character. I borrowed the list below from someone more humorous than I but I endorse them fully:
- Your fences need to be horse-high, pig-tight and bull-strong.
- Keep skunks and bankers and lawyers at a distance.
- Life is simpler when you plow around the stump.
- A bumble bee is considerably faster than a John Deere tractor.
- Words that soak into your ears are whispered...not yelled.
- Do not corner something that you know is meaner than you.
- It don't take a very big person to carry a grudge.
- You cannot unsay a cruel word.
- Every path has a few puddles.
- When you wallow with pigs, expect to get dirty.
- The best sermons are lived, not preached.
- Most of the stuff people worry about ain't never gonna happen anyway.
- Don't judge folks by their relatives. There was no choice there but the company they keep....
- Remember that silence is sometimes the best answer. (This is a really hard one)

Mission 2:
- Conduct research and proposal development efforts for submission against identified marketing targets at TGP.
- Timing has a lot to do with the outcome of a rain dance.
- If you find yourself in a hole, the first thing to do is stop diggin'.
- Sometimes you get, and sometimes you get got.
- The biggest troublemaker watches you from the mirror every mornin'.
- Always drink upstream from the herd.
- Good judgment comes from experience, and a lotta that comes from bad judgment.
- Lettin" the cat outta the bag is a whole lot easier than puttin" it back in.
- If you get to thinking you're a person of some influence, try ordering someone else's dog around.

Letter-for-letter re-creation of my copy of a letter of commendation for service. My copy of the original is 1968 copy technology and the paper is brown and the text is beginning to blend in with the changing color of the paper.
Carroll Childers, MG, ret

NRDU-V:RGZ:1`gm
1650
Ser 138
2 Jul 1968

From: Chief, U.S. Navy Research and Development Unit, Vietnam
To: Commander, U.S. Naval Weapons Laboratory, Dahlgren, Virginia 22448
Subj: Letter of Commendation for Mr. Carrolll D. Childers

1. Mr. Carrolll D. Childers of the U.S. Naval Weapons Laboratory, Dahlgren, Virginia was assigned for temporary duty to the U.S. Navy Research and Development Unit, Vietnam (NRDU-V) as a representative for the Vietnam Laboratory Assistance Program (VLAP) for the period 17 February 1968 to 16 June 1968.

2. Under this program, Mr. Childers was assigned the tasks of keeping abreast of potential developments in ordnance, boat and aircraft armor configurations, personnel protection, and improved armament systems which might have application to the mission of the U.S. Naval Forces, Vietnam. He was further tasked with analyzing current problem areas in those fields, formulating solutions, providing technical direction to developing agencies and subsequently conducting combat introduction/evaluation of the finished product.

3. As a voluntary participant in this program Mr. Childers provided invaluable assistance on numerous occasions to various units and organizations of in-country Naval Forces. The majority of Mr. Childers' tour was spent in the field under actual combat conditions. His resourcefulness and aggressive pursuit of improved techniques and equipment coupled with a superb professional background resulted in outstanding contributions to units requiring assistance. The personal attention and interest

exhibited by Mr. Childers significantly increased the rapport of NRDU-V with operating forces.

4. Working long hours under adverse conditions typified by the "Tet Offensive" and post-Tet enemy operations in the Capital Military District and IV Corps areas of RVN, Mr. Childers either conducted or assisted in the following projects:
   a. Installation of the MK 19 and MK 20 grenade launchers in assault craft of River Assault Flotilla ONE.
   b. Evaluation of the U.S. Army Limited War Laboratory "Battlefield Illumination Kit" aboard River Patrol Boats (PBRs).
   c. Evaluation of a small arms lubricant developed by the Naval Research Laboratory.
   d. Design of a chain link fence standoff armor suit for ASPB MK 48 turrets.
   e. Evaluation of a Ballistic Canopy for assault craft of River Assault Flotilla ONE.
   f. Evaluation of modified ammunition containers for Swift boats.
   g. Combat introduction of a variety of items of personnel and unit equipment for SEALs.

5. Mr. Childers' enthusiasm, ingenuity, and dedication have measurably enhanced the performance of this organization. His exemplary personal conduct, adaptability, and selflessness reflect great credit upon himself and his parent organization.

C.J. Limerick, JR.

Copy to:
DIRNAVLABS
NOL White Oak
Mr. Childers, NWL

* Author R. Guy Zeller, Navy LT, later in life became VADM *

www.ingramcontent.com/pod-product-compliance
Lightning Source LLC
Chambersburg PA
CBHW071452040426
42444CB00008B/1308